THE ADVENTUROUS GARDENER

CHRISTOPHER LLOYD

Introduction by Fergus Garrett

WEIDENFELD & NICOLSON

Previously published in Great Britain by
Frances Lincoln Limited in 2011

This edition published in 2021 by Weidenfeld & Nicolson
an imprint of The Orion Publishing Group Ltd
Carmelite House, 50 Victoria Embankment
London EC4Y 0DZ

An Hachette UK Company

1 3 5 7 9 10 8 6 4 2

A CIP catalogue record for this book is available
from the British Library.

ISBN (Mass Market Paperback) 978 1 4746 1989 9
ISBN (eBook) 978 1 4746 1990 5

Printed and bound in Great Britain by Clays Ltd, Elcograf, S.p.A

www.orionbooks.co.uk
www.weidenfeldandnicolson.co.uk

CONTENTS

Introduction   9
Preface   13

1   TAKING CARE AND MAKING MORE   17
    Hardwood Cuttings   17
    Aids to Rooting   22
    When Kindness Kills   24
    Unusual Ways with Rose Cuttings   26
    Some Reactions to Cutting Back   30
    Pruning the Hydrangeas   47
    Maintaining Mature Hedging   49
    Top-dressings and Root Pruning   51
    The Art of Compromise   53
    For Autumn, Read Spring   56
    Youth at the Helm   59

2   MAINLY WOODY   63
    After the Elms   63
    The Flowering of Non-flowering Trees   73
    Evergreen Broadleaves: the Few Precious Trees   76
    Among the Crabs   79
    Shrubs in Mixed Borders   82
    Roses Take Their Place   87
    Late-flowering Shrubs   91
    Attending to a Shrubbery   98

3   SEASONS AND SITUATIONS   107
    Small Gardens, Large Trees   107
    Dry Wall Free-for-All   115
    No Mutual Antagonism   119
    Spring Meetings   122
    Experiments in Bedding   128
    The Garden in the Heat   140

4   PEOPLE, PLANS AND PLANTS   145
    Planning a Border   145
    Paths, Plants and People   149
    Another Look at Miss Jekyll   153
    Visitors   161
    With Farrer for Reference   167
    Uneasy Status   170
    Prickles and Spines   172
    These Flower Non-stop   175
    Programme for the Greys   184
    Sorting through Eryngiums   190
    The Pampas at Home and Abroad   195
    Flower of the Sentimentalists   199
    Violets   202
    Bulbs to Plant Green   205
    Naked and Unashamed   209

    Index   214

INTRODUCTION
Fergus Garrett

I first visited Christopher Lloyd's garden at Great Dixter as a young student in 1986. I was twenty years old, inexperienced but full of enthusiasm, fresh from an engaging year in Brighton Parks Department and now in the middle of doing a horticultural degree. Great Dixter and Christopher Lloyd were little known to me. My previous gardening experience had been in another world, and even though I had visited great gardens before, nothing I had seen or read about prepared me for this encounter. Long grass brushed our ankles as our group of horticultural students walked down the uneven York stone path to the crooked porch slumped on its elbows in the middle of the rambling medieval manor house.

Christopher met us on the patchy formal lawn bordering the old timber framed building. He was slightly stooped, with hands cupped behind his back, his clothes worn and threadbare, his tweed jacket darned and patched; he had holes in his shoes. He looked unconventional and awkward, yet there was an easy elegance about him. His two dachshunds, Dahlia and Tulipa, snuffled around us disapprovingly, out for blood given the chance.

Wrapped around the house in a series of compartments, divided and subdivided by buildings, walls and hedges, the garden was a brazen kaleidoscope of colours, joyous, eclectic and a curious intermingling of the formal with the informal. The effect was not just appealing but totally captivating. Christo (as he was known to his friends) gardened within a Lutyens masterpiece, added to and strengthened by his parents, Nathaniel and Daisy Lloyd. Settled in a rhythm by years of constant management, comfortable in old age, gnarled and dripping in lichen, the magnificent architectural setting contained a glorious jumble of plants. We were led round on a journey from plant to plant. Each one was carefully considered. Christo's sharpness in observation was patent, as he pointed out details our

9

eyes would otherwise drift over. Notebook in hand I followed, mesmerised and trying to understand.

Clearly this was a garden that bucked the trends. A place gardened by a man given to somewhat revolutionary gestures, yet it had warmth and domestic intimacy (after all, it was Christo's home). Here was a garden that was relaxed in its own identity, where adventurous and inquisitive gardening ruled, and also where plants and people were loved for their own sake. With the barriers of good taste cast aside, Great Dixter had become a place of pilgrimage for everyone who appreciated originality and creativity.

Christo was curious. He had a capacity to focus and concentrate, coupled with an extraordinary talent for communicating. He tirelessly fixed our attention on one plant and then another – the tour took two hours, but could easily have stretched to four. I was intrigued by him and his garden. The visit left its mark and unbeknown to me a gardening seed was sown – a seed that is now a fully flourishing plant.

As I look back to that moment it is clear that Christo inspired me at an important stage of my life. This of course didn't happen over just one meeting. Out of curiosity, on my return to college I decided to read his books. His prose was lively, amusing, opinionated and full of personality, often taking his garden at home as a starting point for literary deliberation. Christo wrote in an accessible and relaxed manner. I was engaged and before long entranced. He shaped my thoughts and doubtless the thoughts of many others like me.

We kept in touch and in time became close friends, eventually working together: I was his head gardener from 1992 until his death in 2006. Throughout the years he has remained the inspirational force in my gardening life. Christo was an individualist who recognised that the basis for good gardening must always be a love of plants and when he found this love in others it formed the basis for easy communication. Throughout his life his gardening was a continuous programme of aesthetic appraisal, considering and reconsidering, refining and reworking, and constantly changing the picture. Christo chose plants for boldness but he was also sensitive and sympathetic to the carefully contrived informality of Dixter. He considered the wider landscape and was curious about human nature. He was also very aware that his exciting style of gardening was intensive and that this could be a limitation – but for him it was the only way to garden. Under him Great Dixter fired on all cylinders, revelling in a no holds barred style, uniting a love of bright colour, imagination, and plantsmanship.

Christo gardened with energy right into his old age. He challenged fashion, flouted conventions and poked fun at correctness. He delighted in experiment and had many successes but equally the road was littered with failures. His experiences were chronicled in his innumerable articles and books. His sound advice was always based on experience: he was a practical gardener first and foremost. His words were driven by actions and his actions fuelled by curiosity. Just as he created one of the most experimental and constantly changing gardens of our time, so he was the best informed, liveliest, and most innovative of gardening writers. He wrote as he gardened, adventurous to the end, with a deep knowledge of plants and practical skills, his thoughts forever conveyed in that distinguished, easy style. His legacy lives on his books: his written word continues to be inspiration to all of us who aspire to garden with something approaching his own imagination, curiosity, and freedom.

PREFACE

In gardening you can never say, 'By now I know all I need to,' and I hope you'll never feel, 'I'm too old to learn new tricks.' For myself, I'm constantly amazed first that gardening technology should still be in as exciting a state of flux as ever in the past (I have belonged almost since its inception to an organization for professionals called the International Plant Propagators' Society, and its annual conferences are enormously stimulating and lively); second, that my own opinions and values are in constant need of revision. Never was there a more foolish saying than that it is a woman's privilege to change her mind.

This book discusses some of my preoccupations and prejudices and it presents some sort of a viewpoint. Readers will often disagree with my value judgements but I hope not resentfully. To be roused into an argumentative frame of mind is in itself no bad thing.

Michael Dover, my editor, who has not only read my manuscript but seems to have extracted its essence better than I could analyse it myself, suggests that I should 'explore my laissez-faire approach' to my subject in this preface, before plunging the reader headlong into practicalities. I think by 'laissez-faire' he's not suggesting indifference on my part. Obviously I care intensely but perhaps I have developed a capacity for not worrying when worry will only take the pleasure out of gardening. We must have some sort of a conscience about the welfare of our gardens but beyond that worry is inclined to lead you into a hopeless frame of mind where you feel that you're never going to cope.

That should seldom be necessary. Of course it can happen that your garden has become too big for you and that your health does not enable you to get on top of a situation. In that case a change in your life style must be faced. But you shouldn't anticipate trouble. As long as you are happy in the present it is better not to look ahead. The future, once we reach middle age, always seems daunting and it'll never take the course you expect it to anyway. Live for here and now, then, if here and now seem good.

Let all your planning ahead be for your plants; a year ahead for annuals, two years ahead for the biennials, an indefinite number of years ahead for the trees. Never take the 'I shan't see it' attitude. By exercising a little vision you will come to realize that the tree, which has a possible future, perhaps a great one, may be more important than yourself, nearing your end. So it's worth thinking more about the tree and giving it a good start in life in the right position than about yourself, except in so far as it is a great delight to see the tree responding and developing under your sympathetic treatment.

Time as a personal factor tends to be greatly exaggerated. It is right to want to fit as much as we can that is worth doing into our lives, but so often the reasons for not growing plants that entail a bit of trouble are unworthy. We refuse to grow these plants not because of other more important commitments but because of other ways in which we're not going to derive value from our time at all. There should be less written about time- and labour-saving ground cover and more about precious plants and how to make them happy.

What else did Michael suggest? Oh yes, not worrying about plant deaths. Well, I don't but many do, taking them, like as not, as a personal insult; and if a label marked 'you are to blame' has to be attached somewhere, it'll probably be on the nurseryman for having sold you a plant that died, even though it was years later.

The best gardening is experimental as well as ephemeral, and both conditions suggest that there'll be deaths somewhere along the line, planned as well as unplanned, or sometimes with a sort of partnership between yourself and nature, the 'merciful release' where you've proffered a helping hand. My friend John Treasure, who lives in a cold Midlands frost valley, lost a vast specimen of *Rosa filipes* 'Kiftsgate' in the last severe winter. It covered such a huge area that replanting the site was like making a new garden. That was in April. A quick decision, his, I mused. 'Was it really dead do you think?' I asked him in July, when he was telling me how other roses had come back. 'Well . . . perhaps not.' I know so well how he felt, and I bet he enjoyed that replanning and planting.

There are some plants that give you intense pleasure throughout your life – at the simplest level, stocks and annual carnations are two such for me; others for which you reach a climax of enthusiasm followed by a slow, relentless decline – *Garrya elliptica,* for me. Take stock, now and again; don't just drift on giving up large areas to plants you're no longer really fond of at all. Deaths can be imposed as well as waited for. You can't always rely on Dame Nature to look your way at the right moment.

Actually it's a great thing to have visitors, whether friends or total strangers, but in either case people who really look at plants and gardens in the same way as you do, to talk to about your garden (and theirs and everyone else's that either of you happen to think of at that moment) as you move slowly, ruminatively, from plant to plant and border to border, looking, thinking, stirring the pot in which ideas and plans were lying dormant. Keen gardeners can be quite starved of the right sort of company. I'm fortunate in that way because many visitors who are appreciative in a constructive way do love visiting Dixter.

'Pruning and planting at the "wrong" times', Michael cited. Well, yes. The wrong time may be the only opportunity and a preferable alternative to not doing something at all. Or it may not be the wrong time, contrary to accepted practice as quoted in gardening literature, if you act cannily. It's all very well to accept received advice and opinions gratefully and at face value, when you're starting, but we graduate. A leavening of scepticism and independence of thought is healthy. You'll make mistakes but you'll perhaps learn not to mind making them. That's a great release from all sorts of inhibitions.

Another way in which you can, I believe, extend your gardening horizon is by not applying every situation, every plant you see, to yourself and your present circumstances. Envy and egocentricity are unnecessary limitations. If you cannot enjoy a plant seen in another garden because it wouldn't grow for you or would be too large for your garden or is unobtainable in commerce, that's paltry. There are far more good plants around than we can ever hope to grow ourselves, but we can still enjoy seeing them. Sometimes they'll have a relevance to ourselves, sometimes not. Never mind which it is; embrace them all if they deserve it.

Gardening is endlessly fascinating and diverse. Those of us who are irretrievably committed are immensely lucky. I am an enthusiast and I do believe that, numerous as the world's band of gardeners is, there should be more of us. Not just routine but mad keen gardeners. Many lack the opportunity but with others it's only a matter of finding the right person to start them off; someone prepared to communicate and share. This book is an attempt at sharing.

1

TAKING CARE
AND MAKING MORE

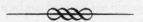

HARDWOOD CUTTINGS

Hardwood cuttings are the easiest kind for the amateur, who has no glass, to take. You cut a branch off a bush, stick it into the ground, and hey presto! a new plant develops. At its easiest, the method is as easy as that. With many willows you can even push a shoot into the ground upside down and it will still make a plant.

With something as easy as a willow, poplar or currant, you should plant your cuttings straight into the final positions where the tree or bush will be required to grow. In most cases, however, you will reserve a special plot as a cutting bed.

When visiting a friend in Norfolk one September I was very impressed by the area in which her gardener was striking hardwood cuttings. This was being done on a small commercial scale; like many others of us, they were trying to make a bit of money on the side to help meet the wage bill. What particularly struck me was, first, the range of material that was being rooted in this way and, second, the fact that most of the cuttings had been made and inserted in early August and many were already rooted. It was, admittedly, a dull damp year, which would be helpful where unprotected cuttings were in question.

The whole point about hardwood cuttings is the cheapness of producing them; the ease for the amateur or part-time professional who does not want the expense of propagation units, heat, glass, mist or any special equipment. The cuttings are rooted in the open air.

17

To enable them to do this without shrivelling up before they have had a chance to make roots, two things are necessary. The cutting material must be sufficiently firm and ripe not to wilt irremediably as soon as it is detached from the parent plant: hence the term hardwood. Second, it must be rooted in a position which receives plenty of sky-light, but no direct sunlight. Sunlight is highly desiccating. Your principal aim, then, is to keep the cuttings alive. So long as they remain alive and healthy the chances are they will make roots in due course.

Often, the best position for your hardwood cuttings bed will be against a north-facing wall. It should not be overhung by trees, making it dark and drippy, and this will rule out most other sunless sites. Alternatively, you can easily rig up a raised bed using a single course of old railway sleepers as the frame and shade it with fine-mesh plastic netting about 3 ft above the bed so that air freely circulates.

The soil must be light and well drained. In my friend's Norfolk garden, the method adopted for ensuring the drainage requirement was to build raised beds some 2 ft high and 2 ft from back to front against a north wall. If your soil is not naturally sandy you can start from scratch with your own cutting compost, such as the John Innes recipe I have quoted many times in many places and shall quote again: 1 part by bulk of (preferably sterilized) loam, 2 parts of peat, 3 parts of coarse sand. The sand will generally be horticultural grit, up to 3/16 in. in diameter. It is ground-up shingle or gravel and obtainable from garden centres or builders' merchants.

The site must be near a water supply, and it should be near to where you or some responsible person frequently passes so that keeping an eye on the cuttings' welfare is almost automatic.

All too often we think about taking hardwood cuttings far too late. It is an advantage if deciduous cuttings – spiraeas, weigelas, philadelphus and deutzias, for instance – still have green leaves on them. The fact that they are still active means that the basal cut will callus over quickly and roots will be made while the weather is still warm. Late cuttings are apt to sit around all through the winter and not start rooting till the spring, by which time the chances of them having died in the meantime from one cause or another are high.

Evergreens, by reason of holding on to their leaves and being thus continuously exposed to the weather, are even more susceptible to winter damage. If you can get them rooted while it is warm they will have a far greater resistance to winter's tribulations. I am not suggesting that, having got them rooted in the autumn, you should in any way disturb them then. Certainly not. Potting them off or lining them out should be left till the spring at the earliest. Slower cases, like *Elaeagnus pungens* 'Maculata' or any of the hollies, will need to wait till June or July.

Although it is an advantage to make cuttings of leafy material, the leaves (whether evergreen or deciduous) should be hard and mature. Many shrubs are still making young extension growth in late summer and early autumn. This, in preparing your cuttings, should be shortened from the tip back to a leaf or pair of leaves of the tough, mature quality that you are looking for. Each cutting will need only two, three or, at the most, four leaves on it, and if these are large, as they would be on the elaeagnus or holly, again, or on a weigela shoot, shorten them back by half with a cross-cut. A mahonia or rose leaf, being pinnate, would have its terminal leaflets removed except for the bottom pair or two pairs. In halving a tough leaf, the one-sided, replaceable blade from a do-it-yourself (Stanley) knife, operating against a hard surface, is effective. With soft leaves, like the weigela's, it may be more effective to fold one or a pair lengthwise, twice, and then cut across, half-way up, with a knife (a very good way, this, with the voluminous but pliable hydrangea leaf).

Even hardwood cuttings will probably be only 4 in. or so long (not the 10 or 12 in. so often recommended in gardening literature), and the bottom two-thirds of the cutting has its leaves cleanly removed close to the stem. In most cases you will do this with a knife or blade, but with some cuttings like hebes or *Euphorbia characias* you can pull the leaves off and they leave a clean scar without tearing.

We now come to the basal cut itself. This is usually made immediately below a node. Whether it slopes or is made straight across matters little. Perhaps the larger exposed surface of a sloping cut gives greater scope for the development of root initials, and if you are using a knife a sloping cut is the only sort that can be made tidily. The easiest way in many cases when the wood is thick and hard (e.g. mahonia, rose) is to cut first with a pair of scissor-action secateurs (not anvil-type, as these bruise the wood against the anvil) and then, with a blade, remove a further thin shaving, cutting down on to a hard surface.

Never allow the exposed surface of your cut to dry. As a matter of routine I immediately dip it into a jar of water. When I have made my batch and am about to insert them (or stick them, as the Americans say), I dip the moist end of each into a rooting powder. For hardwood cuttings use the strongest formula.

When inserting your cuttings make sure, first, that the bed into which they are going is firm. It is difficult to describe exactly what I mean by this. It doesn't want to be packed dourly tight and hard, but it doesn't want to be spongy either. Some propagators push their cuttings in, to the required depth, which is so that the lowest leaf remains just clear of the soil. In this way they ensure that the bottom of the cutting does not have an air space below it. The other method is to make a hole with a dibber and drop the cutting into this, but you then have to gauge the exact length of hole needed (a) to bring the lowest leaf into the correct position and (b) to bring the bottom of the cutting to the bottom of the hole leaving no air space between the two. I usually prefer the latter method. You soon learn by eye and experience what depth to dibble to. The trouble with the first method is that you rub much of the hormone off the base of the cutting as you push it through the soil. But then if you pasted a lot of powder on in the first place this would not matter. If the compost has been too firmly compacted, pushing the cutting in may well damage its rind at the cutting's base.

Next you water, and there you are. A dull, muggy day is best for the job, because with a lot of cuttings lying around in transit the task of preventing them from drying is difficult on a bright day with a brisk breeze. Evergreens are easily dehydrated without showing it. Polythene bags are a boon, of course, though not so good when you are dealing with prickly subjects like roses. Even a hardwood rose cutting, incidentally, need consist of no more than two nodes, the upper with its leaf still attached, though shortened. Thus the entire stem of such a cutting will be submerged, leaving only the leaf and 'eye' (which is the axillary bud) visible. Don't worry about heels or anything like that. A nodal cutting made of firm wood is entirely satisfactory.

Turning to a few examples: honeysuckles should be done before their leaves yellow. Take trails of the current season's growth and chop them up so that there is a pair of leaves at the top (which you can halve) and a node at the base from which the leaves have been removed. Young shoots of bush honeysuckles like the winter-flowering *Lonicera* × *purpusii* are also suitable. Here I leave two shortened pairs of leaves, since the nodes are quite close along the stem.

Many hedging plants respond well: yew, holly, box, bay laurel, escallonias, rosemary, lavender. I have made lavender cuttings, hundreds at a go (you can push seventy into a seed tray if you want them to remain mobile) at many seasons up to January, but have had by far the highest take from those struck in August and early September. It makes all the difference, too, if the stock plants are themselves young (two years old is ideal) and strong; not those old runts at their last gasp that you're planning to get rid of. Perhaps you'll find a mass of ideal young shoots on young bushes in a friend's garden. Assure him from me that their removal will do the milch cow no harm at all. On the contrary she will, as is well known of cows, feel all the better for the milking.

As long as you are not too fearfully exposed to north and east winds, why not be different with a hedge of *Griselinia littoralis*? Some daphnes are easy, others less so, and some swinish. The easy ones are *Daphne odora* and *D. pontica* (delicious scent on May nights), while it will be worth having a bash at *D. tangutica* and *D.t.* Retusa Group, although berries are really a welcome let-out. Hebes can be rooted at any season, but the whipcord types with their tiny adpressed scale leaves so amazingly resembling conifers are less easy than the rest and this is the time to catch them.

If propagating your caryopteris immediately after flowering, take the entire length of the current year's stems and remove the top half which has borne the flowers. You can do this while they are actually flowering if (like me) you have a hard heart. Somewhat tender evergreens like *Myrtus communis*, the common myrtle, would really be safer under cold glass in the winter, but if you do not have this, don't be deterred. You may be lucky and the winter mild. Always back your luck.

Skimmias root splendidly from cuttings of their new shoots trimmed where they started from the previous year's, and this is the only way in which you can be sure of your sexes. *Osmanthus delavayi* can be rooted from young shoots (remove the soft tips) in late September, but will certainly take a year about it. You can see when they've rooted because each rooted cutting will put out a new young shoot. As they root unevenly, the best way is to lift a batch, pot those that have rooted individually (or line them out in a shady place) and put back the rest. All sorts of dwarf and bushy conifers: junipers, cypresses, thujas are suitable. Catch your *Pittosporum tenuifolium* and *Choisya ternata* in between flushes of growth; you can never be sure that their growing season is at an end. And have a go at some of those fancy brooms (*Cytisus*) that will not come true from seed. All the olearias: such a fascinatingly diverse group but again you will, with many of them, have to hope for a mild winter.

You might even have a shot with camellias. Struck in early August they will probably have rooted by the following July. Again, choose young and vigorous material if you possibly can, and rub out all flower buds as they appear.

AIDS TO ROOTING

The striking of cuttings, as I have pointed out before, is a race between rooting and rotting. If you can prevent a cutting from rotting for long enough, it will probably root in the end, sometimes sooner, sometimes later. Obviously the sooner it roots, the sooner will the battle against rotting have been won.

Rotting is most likely to occur in the early stages of taking a soft or half-ripe cutting, when it has to be kept in a close, unventilated atmosphere so as to prevent it from wilting before it has had a chance to root. Fungus spores can easily gain control under these ideal conditions, and the most likely culprit is *Botrytis cinerea,* the grey mould fungus, which mantles its victims like a fur coat. It operates far more easily on soft tissues than on leathery. Choisyas and skimmias are leather-leaved and safe enough, but caryopteris and erysimum are soft and easily assailed. So are woolly-leaved plants that hold the moisture: *Senecio cineraria,* lavender, gazanias – such as these. Also wax-coated leaves that lose their protective wax: some acaenas, rues, dianthus. In these cases it may be possible and hence wiser to allow ventilation almost or right from the start.

Another way to protect your cuttings against rotting is to spray them with a protective fungicide every week while they are in the close frame. If you vary the fungicide on a rotational basis, you prevent resistant fungus strains from developing. Other things being equal (by which I mean the moisture factor and the keeping of moulds in abeyance), a close atmosphere with high humidity helps to speed up rooting and it also retains heat in a frame, which again speeds things. Nowadays, most professional propagators use mist units with bottom heat. In this way you can keep your cuttings permanently moist without cutting down on ventilation, and this minimizes fungal dangers, especially as a fungicide is included in the mist. The bottom heat keeps the cuttings' toes warm and encourages root formation. However, although many amateurs have played with mist propagation units, I don't think they are really suitable or appropriate for small-scale work.

Another way to speed up rooting and foil the rots is by the use of synthesized growth-promoting substances similar to the rooting hormones produced by the cutting itself. These can be either in liquid or in talc powder form. The latter is the more convenient to use, I find. After making your cutting you dip its base in water, shake it, dip it in the powder, tap off excess, and then insert the cutting in the rooting medium.

22

It is quickly done but it is 'one more thing'. Is it worth it? That is the question, and at one time I was saying no; I could find no difference between the speed and profusion of root formation on treated and untreated cuttings. However, growth substances are widely used by professional propagators, so I had to keep an open mind (always an unsettling state to be in).

A chance remark from a friend at Kew (rather like saying a friend at Court, in gardening parlance) put matters in an altogether different light. You must use these synthetic hormones fresh, he told me. I realized that, being of an economical disposition and disinclined to throw anything away without good reason, I'd been using the same canisters of hormone powder for (dare I admit it?) twenty or twenty-five years! I do hope that some readers have been similarly guilty of this practice.

Anyway, I sent off for some fresh and I must say that the results in many cuttings that I regularly take each year were spectacular: *Escallonia* 'Iveyi' (usually making a huge lump of callus but roots only very costively), choisya, *Cytisus* 'Hollandia', *Cistus* × *cyprius, Helichrysum splendidum, Mahonia japonica, Phlomis fruticosa,* to name a few examples. Others, such as evergreen ceanothus, *Elaeagnus pungens* 'Maculata' and *Osmanthus delavayi*, were less successful. There are many possible reasons. I may have used the substance at the wrong strength. More likely still, I was taking the wrong sort of cutting material, either from too old a bush or in the wrong condition of ripeness.

These synthetic hormones are marketed by May & Baker in three strengths. The weakest is used for soft cuttings. As soft cuttings root extremely easily and quickly if they root at all, I don't bother with this. The other two strengths are for half-ripe and hardwood cuttings. If you can't make up your mind whether your cuttings are in the one state or the other I advise settling for the stronger brew. Murphy offers one rooting compound only, to cover all types of cutting, which sounds nice and simple but we all know what comes of being all things to all men. Remember that these synthetic hormones gradually lose their strength and should not be relied upon after two years. I checked with the makers about this. I also suggested to them that it would be a good idea to print on each container the date by which its contents should be used, as is done for photographic film and for some foods. We could then tell at a glance whether what we were buying was fresh. My suggestion has not been taken up, you will regret to learn.

WHEN KINDNESS KILLS

The enthusiastic amateur gardener has a tendency to get busy with certain jobs before it is wise to do so. A notorious case in point which I frequently deplore is the premature sowing of seeds in spring. A similar but much less obvious situation arises in the autumn, when many rooted cuttings are disturbed and destroyed.

Here, in truth, is a great temptation. You have had success in rooting cuttings and this enthuses you – there is nothing more exciting in the whole of gardening than successful propagation, I should say. As a good father or mother to your new children you wish them to lack for nothing, so you turn them out of their cutting compost, which contains little or no nutrient, and pot them individually into something nutritious.

This is all very well in summer, when plenty of the growing season remains, but from the end of August onwards, as the light fails and temperatures fall, it results in very heavy casualties. By then, even as easy a case to deal with as a hydrangea cutting will be better left alone till the following March, if only recently rooted. I have lost many young hydrangea plants in the past (especially *Hydrangea paniculata* types) by potting them off singly in autumn. All goes well at first, but when spring comes round their dormant buds fail to break. They have died, and the fact that the pot is full of healthy roots is to no avail. Cuttings of the trickier *Elaeagnus pungens* 'Maculata' will probably not have rooted before November. If potted off singly then, the whole lot will succumb, but left till March they will survive.

When I attend International Plant Propagators' Society conferences, speakers often make this point. Apropos of Japanese maples and the rooting of *Acer palmatum* cultivars from cuttings taken in June, it is recommended that, when rooted, they should be left in their boxes over winter and potted in spring. A speaker who runs a large nursery of container-grown plants told us that none of any variety of shrub are potted after the end of August.

Here follow some examples of shrubs that from my personal experience I have found to be best left alone when they have rooted in autumn. Like the contents of a rag bag they have no pattern or sequence: hollies; brooms; clematis; *Convolvulus cneorum*; rock phloxes; polystichums of the kind you root from bulbils along the frond midrib; *Syringa vulgaris* cvs. (common lilac); magnolias; camellias; rhododendrons and azaleas; daphnes; rosemary and lavender; buddleias; viburnums; summer jasmine; roses; *Sambucus racemosa*

'Plumosa Aurea'; all kinds of ceanothus; conifers; cistus; *Solanum crispum* and *S. laxum* (syn. *S. jasminoides*); *Euphorbia characias* and subsp. *wulfenii;* myrtle. Dendromecon, garrya and some eucryphias, notably *Eucryphia* × *nymansensis* 'Nymansay', are so touchy about root disturbance that they should be struck individually in Jiffy No. 7s in autumn (October is a good moment for garryas) and potted when their roots become visible the next spring.

In holding them through the winter, keep the temperature low. Then they won't be using up energy and it won't matter that there is little or no nutrient in their cutting compost. There is the alternative of higher temperatures and supplementary lighting, which will prevent even a deciduous azalea or acer from shedding its leaves and becoming dormant in the winter, but I mention this only in passing as few amateurs are likely to want to experiment with these exacting techniques.

Turning now to tender bedding perennials. Each year I raise a fair-sized batch of *Felicia pappei*, which carries bright blue daisies over a long summer–autumn period on a bush of fresh green colouring. My one-time method of propagation was to take cuttings in September and pot them individually the next month as soon as they had rooted. They are overwintered under glass from which frost is only just excluded, and all would go well till February, when the tips of every leaf turned brown, the dead area gradually working back. I used to think we had let the frost in, without knowing it, but since I have in more recent years struck the cuttings a month later and left them in the pots they were rooted in – about ten to a 3-in. pot – right through the winter, the trouble has completely disappeared. Now I do the same for all my autumn-struck soft cuttings, and the saving in greenhouse bench space is enormous. By the time the plants are given individual treatment in late March or early April, they can go straight into unheated frames.

Included in this category are fuchsias, gazanias, osteospermums, bedding verbenas and pelargoniums ('geraniums'), *Argyranthemum frutescens, A. foeniculaceum* and *Tanacetum ptarmiciflorum* (syn. *Chrysanthemum ptarmiciflorum*); *Convolvulus sabatius* (syn. *C. mauritanicus*), *Mimulus glutinosus, Cuphea ignea* and *C. cyanea, Centaurea gymnocarpa, Senecio viravira* (syn. *S. leucostachys*), *Helichrysum petiolare* cvs, *Plecostachys serpyllifolia* (syn. *Helichrysum microphyllum*) and named bedding penstemons. Some gardeners will prefer to root these kinds of plants quite early, say in August or early September, and keep them moving all through the winter at a minimum temperature around 45° F. To be worth the great extra expense, there would need to be some special advantage in having large plants early in the following season. My way is not merely more economical in heating but ten times more so in space (this being the number of cuttings I get into one 3½-in. pot).

Yet another method that will recommend itself if you anyway have the heat available and paid for by the public (I refer to establishments whose bills are footed by the public in one way or another), is to grow on stock plants through the winter and take cuttings from them in February. But most of us are counting our pennies.

There are some rooted cuttings that I *do* pot off in the autumn, it still seems to me with advantage. These include *Ballota pseudodictamnus*, *Lithodora* (syn. *Lithospermum*) 'Heavenly Blue', helianthemums, acaenas, *Anthemis punctata* subsp. *cupaniana*, *Scrophularia auriculata* 'Variegata', easy rock campanulas such as *Campanula portenschlagiana* (*muralis*) and *C. poscharskyana*, thymes, aubrietas, *Mahonia japonica*, and *Senecio cineraria*.

All, be it noted, are evergreen or wintergreen (i.e. with a winter growing season), and all are hardy or nearly so. With no more protection than that afforded by a cold frame they are capable of growing right through the winter. They benefit from individual treatment in a nourishing potting compost. But they are the exceptions, make no mistake.

UNUSUAL WAYS WITH ROSE CUTTINGS

We keep learning, and it is chastening to be reminded how little we yet know. You'd have thought, for instance, that if all forms of clematis could be easily propagated from cuttings that's how they'd have been done right from the time, more than a century ago, when they first became a popular florist's flower. Not a bit of it. Only in recent years has the major swing-over taken place from the propagation of all large-flowered and many smaller clematis hybrids and species by grafting. Grafting still has its advocates but at least nine-tenths are now done from cuttings, usually internodal cuttings, and no more sophisticated equipment is needed for the purpose than was ever available.

With roses, grafting is still (except for the miniatures) the principal commercial practice. The entire organization of a rose nursery, its machinery, equipment and labour force, are geared to rose production by bud-grafting and it would be such an upheaval to change over to the very different technique of raising them from cuttings (or from tissue cultures) that this is unlikely to happen in the near future. But it could happen, all the same, and especially now that it is understood that roses themselves can be propagated by the very economical method of internodal, or single-node, cuttings.

The traditional method for preparing a stem cutting is by making the bottom cut just below a node, the top cut (unless the tip of the shoot is left intact) above a node. Thus, at least two nodes are included in every cutting and this is potentially extravagant in plant material. It is a principal reason why cuttings have never been much used as a commercial method for propagating roses, seeing that only one node is needed for budding.

I first learned about the single-node way of taking rose cuttings at a conference of the International Plant Propagators' Society which I attended some years ago now (1970) at Nottingham University's School of Agriculture, where Dr Elizabeth Marston had been carrying out experiments.

To make a single-node rose cutting, all you have to do is to cut with secateurs (those having a scissor action, not the anvil type) half an inch above a node and two inches or less, if necessary, below it. A surprising point arising here is the fact that exposing soft internodal pith at the cutting's base does not induce rotting. You insert your cuttings in a suitable rooting medium – say half peat and half grit – and wait for results. If, as in the Nottingham University experiment, the cuttings go under mist and are given bottom heat of 65⁰ F, they root quickly. But one conference member told me that he had succeeded by inserting them straight into the open ground under polythene tunnels (such as are extensively used nowadays for forcing strawberries). Twice-daily hand-watering from a can was the only attention required.

In the actual experiment, whose results I witnessed two years later, an entire plant of the HT rose 'Prima Ballerina' was chopped up and used for cutting material, some wood being ripe, some half-ripe and some soft. This was done on 8 August and the plant made some 130 cuttings. Two years later there were sixty-nine plants, nearly all of them sturdy bushes, in the experimental plot. On getting home I forthwith tried the method out for myself. On 29 July I made one seedbox-full of thirty cuttings of' 'Peace'; another divided among three Floribunda cultivars: twelve of 'Allotria', thirteen of 'Europeana' and seventeen of 'Red Wonder'. They went into my cold frame with its double walls and double glazing.

By the end of August, some of the Floribundas were showing roots through the bottom of the box. A point I noticed was that, although each cutting had only one leaf, it was very large and horizontally space-consuming. Where one leaf overlapped another, the lower one yellowed and decayed prematurely. I should have shortened each leaf at least by its terminal leaflet, if not also the pair behind that.

I did not disturb the cuttings that autumn, the season being well advanced. They overwintered in a ventilated cold frame and many of them retained their foliage throughout. When I dealt with them, potting them off individually at the end of March, twenty-six 'Peace' had rooted, six 'Allotria', ten 'Europeana' and nine 'Red Wonder'. Not a startling result but sufficiently encouraging. If you took your cuttings a month earlier than I did and had them well rooted by early August, they could be potted individually then, instead of waiting till the next spring.

Internodal rose cuttings have many possibilities and implications. As private gardeners growing roses for fun, we can now ask our friends for cutting material without plunging deep into their bushes in search of heels. Commercially this economical method of getting roses on their own roots could be a boon to the container trade at garden centres. Grafted HT and Floribunda roses are stunted and uncharacteristic when their roots are confined in even quite large containers. Yet the container trade has a tremendous outlet for the sale of roses in their flowering season, when the impulse buyer can not only see what he wants but carry it off with him on the spot. Own-root plants might turn out to be the answer. Little is known about the performance of different varieties when grown this way, but then singularly little is known about the performance of different varieties when grafted on different rootstocks. These rootstocks are nearly all seedlings, and every one of them is genetically different from the next and thus establishes a different relationship with the scion grafted on to it. On its own roots, the intermediate imponderables of the rootstock will be eliminated. And so will its suckers.

Suckers lead me to the other relevant subject of Dr Marston's experiments, which was rose propagation from root cuttings. The necessary ingredient for success in propagation by this method is that a detached piece of root shall be capable, when given its independence, of making a shoot bud. Shoots are normally produced by stems, not by roots, so the latter capacity is not to be taken for granted.

Roses have it, as we always knew, for how else should their rootstocks sucker, as they all too frequently do? I recently had a striking example of suckering by rose roots in my own garden. I had planted the ultra-vigorous

'Paul's Himalayan Musk' Rambler to go through a Judas tree. I had struck this rose from a cutting off one in Maurice Mason's garden, so I knew it was on its own roots. The Judas tree got blown to bits in a gale and had to go, and so did the rose, since it had lost its support. After its removal the whole area, up to a distance of 16 ft from where the rose had been extracted, became riddled with suckers from the broken roots we had left behind. Yet until then the plant had shown no inclination or tendency to sucker at all.

Dr Marston showed that even the more specialized and man-made HT rose can be propagated from root cuttings if the stock plant is growing on its own roots (quite a big 'if'; the stock plant is only likely to be growing on its own roots if it was one you propagated from a cutting yourself). In her experiment, 2-in. long root cuttings were taken in January and laid horizontally in a peat and grit mixture. They were of all thicknesses, but the thinnest produced no shoots while the thickest produced shoots prematurely, before new roots had developed to support them, and these shoots withered. (In such a case it would be wise to anticipate trouble and make stem cuttings of these shoots while they were in good condition.)

Rather surprisingly (to me), it was found that horizontally orientated cuttings gave better results than cuttings inserted vertically, and, of course, they tax the intelligence less because there's no need to remember which is top and which is bottom. No growth substances were used, but the cuttings were gently shaken in a polythene bag with 15 per cent captan dust, as a protection against fungal infection.

Some cuttings were placed in a warm greenhouse with bottom heat, and these developed quickly. Others in a cool greenhouse from which frost was only just excluded took some months to regenerate, but the final take was as good. Cuttings taken in December and January gave better results than later batches in April.

Not one but a number of shoots are likely to develop from a root, so your young shrub might have the appearance of suckering instead of growing on the normal leg. I cannot see that this should matter. Varieties used successfully include 'Fragrant Cloud' and the repulsive but ebullient 'Prima Ballerina'.

The young plants were potted off into 3½-in. pots and developed very quickly, but the first flowers borne that summer were thin or even single and not characteristic. To build up a strong plant it would be wise not to allow flowering at this early age, but to disbud, as you would a stem cutting.

One disadvantage in root cuttings is that you are virtually obliged to sacrifice the parent plant. You can't deprive it of its roots and then expect to keep it as a thing of beauty and a joy for ever. Probably the method has no general application, but it is none the less intriguing and open to any of us to try who already grow roses on their own roots.

SOME REACTIONS TO CUTTING BACK

'There is a psychological difference between cutting back and pruning,' I wrote in *The Well-Tempered Garden.* 'Pruning is supposed to be for the welfare of the tree or shrub; cutting back is for the satisfaction of the cutter. Some gardeners have a cutting back mentality . . .' True as far as it goes. Regular cutting back of shrubs which should appear to be growing freely, their branches laden with swags of blossom, transforms them into a kind of vegetable hedgehog on stilts. You see this all the time in public gardens where there is no one with any knowledge in charge. Forsythias, flowering currants, philadelphus, weigelas and white spiraeas are favourite victims on the altar of hedgehoggery.

But there are many situations in the long-established garden where shrubs have grown so large for their position that they are killing all their neighbours, blocking access to a doorway or a passage along a path, or taking light from where light is wanted – a window, for instance. If, by cutting hard back, you can reduce them to acceptable dimensions for four or five years, you may have hit upon a happier solution than the alternative of grubbing. Same thing when a shrub has become scruffy, with bald patches like a bird in full moult.

What we should like to know, however, is whether the cutting back treatment is likely to be successful. Because if not, we might just as well take a short cut to the inevitable. Instead of waiting in vain for a mutilated shrub to show renewed vitality (scratching away shavings of bark every week or two and muttering 'it's still green underneath'), we can save ourselves these agonies and grub the plant at the outset.

But how are we to know? None of the evidence has been collected for an answer to this question so there follows a tentative effort on my part to fill a gap. For the most part I shall concentrate my comments on what I have learned from personal experience: it is so easy to go wrong when you start quoting other people who may themselves be quoting yet another 'authority'.

Before itemizing, it is my duty to issue a practical directive. Cutting back into thick old wood will generally require a saw. Make sure the cut is clean. You can always avoid leaving a torn wound where the lopped branch took charge before the cut had been completed, by sawing it in two stages. Make your first cut a foot or so beyond the final cut. The latter can then be achieved with full control. You can easily support the weight of the stump being removed with your non-operating hand. Finally, protect the open scars with a flexible bituminous paint. 'Arbrex' is the best-known trade product at the moment, but there are others as good.

It is wise to hedge your bets. In case the patient dies, root some cuttings (or raise some seedlings) off it first – preferably in the previous year so that you're ready with a replacement.

If I were to be asked the best season for cutting back, I should be hard pressed to say. Obviously it is wrong to cut into anything half-hardy just before or during the winter. But during the winter often seems a good time for really hardy stuff like lilacs. You can see what you're doing when the leaves are off, and anyway you've more time and it's a nice warming job. Whether the lilac would be happier to be massacred at some other season, it hasn't told me. Why can't people who are so keen on talking to their plants these days (as a substitute for looking after them as they should) train the plants to talk back? (They might get a shock.)

The prunus tribe is generally accepted to be best tackled in early spring, when sap is already on the move. It is subject to some dire diseases – bacterial canker and silver leaf, in particular – which are said to be less likely to gain entry if pruning or lopping is delayed till spring. On this I have to accept the advice of fruit-growing experts with a detailed knowledge of plums and cherries from a commercial standpoint.

When the shrub in question flowers in spring or early summer, we shall frequently attempt to eat our cake and have it by enjoying its flowers and then off with its head. Enough of the growing season remains for it to push out some new shoots before winter's arrival.

When you have to cut back into old yew wood, its resilience is such that it will always break into fresh green growth even if your cuts have to go back to its trunk, though it is not quite as simple as that. Say you are taking an overgrown hedge or topiary specimen back from a path, you will make your cuts so that the hedge's new outline is still correct with geometrical planes. We did quite a lot of this one March, and a drought ensued, as so often happens in the spring. A high proportion of the neatly shortened branches died back to the trunk, and it was from the trunk itself that the new growth came. We had a very wet September that year. To my astonishment, many cut branches that had remained completely dormant till then, and had appeared to have died, broke into young growth, right to their tips. Those that didn't and had really died eventually had to be removed. There is nothing unusual about this sequence. When you cut yews and a great many other shrubs (some of which I shall be naming) hard back, they habitually die back further than you intended or really wished.

The interesting point, to me, is the revitalizing effect of prolonged wet weather. It is well known to greenhouse owners that when you cut an old shrub hard back in order to rejuvenate it or to get young cutting material from it – the fuchsia would be a typical example – you make a habit of daily or, preferably, several times daily syringing the dormant old stems in order to induce them to break. I have little doubt in my own mind that if what we call 'growing weather' followed upon the cutting back of a shrub, i.e. if it were living in a kind of Turkish bath atmosphere, all its naked stems would sprout to their very ends. As it is we have to accept that some will, some won't, and that if we're lucky with the weather there'll be greater success than if we're not. We cannot be expected to go tripping round our gardens with a water syringe, five times a day.

Overgrown hedging often causes headaches in the perplexed owner. Hornbeam and beech are comparable in appearance, retaining their dead foliage through the winter. However, whereas you can be as severe as you like on hornbeam, beech will react far more slowly and patchily. There is a beech hedge on stilts at the end of our village, rising above a high brick wall. Its owner had to reduce it somehow, so he halved the width of the hedge by tackling the inner face, behind the wall, first. It broke very slowly but is gradually recovering. The outer half still projects beyond the wall, but could be reduced in due course. It is just as well to take two bites at the cherry where the patient is a bad recoverer and, furthermore, to allow an interval of several years between bites.

Holly (*Ilex aquifolium*) causes no anxieties when cut into old wood but ilex, the holm oak (*Quercus ilex*), simply will not have it. This makes an excellent large hedge and needs clipping just once a year, but if you take it back into bare wood you'll get a nil response. Indeed, oaks as a tribe respond poorly to such treatment, whereas most native woodlanders can be hit back to the ground, if you so wish: stooling, this is called and I shall return to the practice.

The commonest coniferous hedging materials are *Thuja plicata* (western red cedar), *Chamaecyparis lawsoniana* (Lawson cypress), *Cupressus macrocarpa* (called macrocarpa for short), and the fashionable Leyland cypress which is a bigeneric hybrid and has the greatest vigour of all evergreen hedging. This is × *Cuprocyparis leylandii*.

I was once so incautious as to write in an article that *Thuja plicata* does not break from old wood, nor *Lonicera nitida* – you must start again from scratch if your hedge has got out of hand – but that Lawson and Leyland cypress do break. No sooner written than I was tormented with doubts, especially about thuja and Lawson, whether I hadn't got them back to front. So I wrote to John Street, whom I knew, because he had dealt with just this point in a broadcast we made together years ago. 'I'm dreadfully sorry,' he replied, 'but you're out of luck! It's the other way round. You can cut back *Thuja plicata* but not Lawson Cypress! However, there is a ray of hope. The Lawsons can be cut back if they are green below. That is to say, if they are still furnished at the base, you can cut them even to about 6 ft and they will grow away happily from the top. Unfortunately, though, they will not break from a bare stem.'

Meanwhile, having read my printed balderdash, Mr Arthur Gould wrote in confirming John Street's definition and 'taking issue' with mine: 'In my experience one of the few defects of Lawson Cypress is that it will not generally break from old wood whereas *Thuja plicata* is most accommodating in this respect.' He then delivered a further body blow: '*Lonicera nitida* I regard as being almost impossible to treat too severely. I inadvertently set fire to a length of this hedge last year and all that remained were a few charred stumps. In a little over 12 months these have regrown to give a reasonably presentable hedge about 18 in. high and through.' At least I can take advantage of my hard-won experience, I thought. A specimen of *L. nitida* 'Baggesen's Gold' on a prominent corner of my Long Border had, in the course of time, become overgrown, although I clipped it annually in February with secateurs. So I sawed it back to stumps. It responded beautifully with young shoots in the ensuing growing season. But there was a pay-off, for in the following year first one shoot and then another and then the whole bush died. So I had lost all round on that exchange. I hope the reader's course is now clear.

It is well known that if you cut *Cupressus macrocarpa* back into bare wood, it dies, inexorably. Leyland is the hybrid between this and *Xanthocyparis* (syn. *Chamaecyparis*) *nootkatensis*. I asked Mr Jobling, of the Forestry Commission's research station at Alice Holt, Surrey, about this one, and he said that you could cut Leyland back into, say, four-year-old wood without ill effects but he was not dead certain whether it would break from really old, bare wood. His general comment on this hybrid was interesting. 'The thing about Leyland is its massive sideways growth. I think gardeners are going to be sickened by it,' he said. Take heed all. Leyland is fashionable because of its phenomenal growth rate but it is an overbearing servant.

The true laurel, otherwise known as bay, can be cut into very old wood without ill effects. Indeed this becomes essential whenever it has been mauled by a specially severe winter. The old wood will break along its entire surface with young purple shoots, but this may not happen till late June or July. Myrtle, another dicey subject from the Mediterranean, is an absolute marvel in this respect. Our plant is now 10 ft high by 18 ft across, but had to be cut right back to ground level after the 1962-3 winter, and also on two previous occasions in my lifetime. It might be said that a hard winter every so often is good for it.

The Australian bottlebrush tribe of callistemon varies a good deal in hardiness among its members (it belongs to the *Myrtaceae*), but if they'll live with you at all – and many are hardier than they are given a chance to prove – they will certainly respond well to cutting back. My *C. subulatus* was taken back to the ground in 1963, not that it had been killed that far but its wood had been split by frost. By 1975 it was so large and unruly again that I was forced to take action without winter's help. I cut it pretty hard all over in June. It flowers in July, but by June I could see that this was to be one of its shy-flowering years (all depends on the heat of the previous summer) and I wanted to give it as much recovering time as possible in the current growing season. In fact it took ages to respond but eventually did, right up to my cuts. However, it did not make nearly enough growth that summer and autumn to flower the next year – 1976 – but in that year it made giant strides and fully furnished any nakedness of exposed wood.

Eucalyptus is a third member of the warm-temperate and subtropical myrtle family. Some eucalypts respond so well to cutting back that you can treat them as shrubs by stooling them annually. *E. gunnii* is one of the prettiest for this treatment. It will produce nothing but small, rounded, juvenile foliage of brilliant glaucous colouring, and this can be contrasted in a mixed border or shrubbery with purple-leaved *Cotinus coggygria* and various berberis of the same persuasion.

Certain eucalypts have to be treated with greater circumspection. *E. perriniana* is especially noted for its glamorous perfoliate juvenile leaves which surround the stems in a chain of discs, but it sulks badly when cut back for foliage effect and can easily be killed. However, if you want a bush with round leaves you don't want a tree with long thin leaves, which is the alternative. So I do in fact cut mine back to a 5-ft framework, but not everything in one year. Some branches are left intact in one year and shortened the next. It is sensible, however, to consider this species as a short-term proposition and to replace it fairly frequently. After all, the gums grow so fast that there's little to lose.

The pomegranate, *Punica granatum,* is a very long-lived shrub of far greater hardiness than you might think, though it deserves a baking wall or corner if you are to enjoy as many of its scarlet blooms as possible. I have often seen this cut hard back, either by winter or by humans, and it invariably responds with alacrity, but it is the unpruned specimen that will carry the blossom.

Hardy hibiscus, usually of the species *H. syriacus* but also *H. sinosyriacus,* are in much the same case as the pomegranate except that they would flower on their young wood, following regular spring pruning, if our summers were hotter. As it is they must be left as large and ungainly unpruned shrubs (they need no wall protection) if we want them to flower regularly, but if they have to be cut hard every now and again they'll come back splendidly.

Annual trimming of the young shoots on a shrub is a very different matter from cutting into older wood. Most keen heather owners will go over them with shears in the spring, but suppose they had an old heather planting that had become scrawny and shy-flowering with only short spikes at best? Can it be rejuvenated by cutting to the bone? I decided to try, on an ancient patch of *Erica carnea* 'King George'. After all, I told myself, the burning of heather for the sake of grouse is expressly aimed at lots of tender, succulent young growth to fatten the birds. And anyway I'm not very fond of heathers and didn't mind if mine died. They died. Not a wisp of life returned except for some self-sown *E. erigena* (syn. *E. mediterranea*) seedlings from a nearby bush. I then realized that the regeneration of heather on grouse moors is from seedlings, not from the old burnt bushes. You cannot cut heathers back into wood that is more than three years old and get away with it, an expert subsequently informed me.

Evergreen ceanothus often run into trouble. They are generally planted against a wall, for protection, but they are such vigorous growers that in no time they are across your windows or pushing forwards across a path. You can train a ceanothus to fill an allotted wall area and then make a practice of shearing it over annually, as soon as flowering is completed, say in early June. This is never as wholly satisfactory and exciting in its results as the unpruned bush that has been allowed to have its head, but it is better than no ceanothus, perhaps. What you can seldom get away with is cutting a ceanothus back into wood that is more than one year old. Often the owner will cut the lower, forward-growing branches hard back while leaving the top, upward-growing shoots untouched. All the bottom becomes bare, with dead snags for decoration, while the top, in rude health, bulges in unbalanced obesity. In fact, you've chosen the wrong shrub for the position. Substitute something with pliable and easily trained branches like *Abutilon megapotamicum, Jasminum nudiflorum,* a peach or a morello cherry; or with branches that respond well to being spurred back like *Chaenomeles speciosa,* the Japanese quince, or a fruiting pear.

Having written the above I must, in honesty, confuse the issue. I have sometimes cut large *top* branches from *Ceanothus dentatus* var. *floribundus* and also from my ancient *C. impressus* (planted 1950; see *The Well-Tempered Garden*) and have been rewarded with some young and vigorous growth from just below the cuts. So it's worth a try if you're desperate, but on the top rather than the sides of a bush.

The *Cytisus* group of brooms is similarly touchy. Even in shortening its flowered shoots by half, I killed my *C. × kewensis.* Certain other brooms will sprout like pollarded willows, however. The spiny *Genista hispanica* is best restored to comeliness every few years by regularly reducing it to stumps (it makes a neat and effective low hedge where you're anxious to stop people taking short cuts), and the not dissimilar gorse (*Ulex europaeus*) is far showier a year or two after this treatment.

Cistuses resent being cut into old wood. They often look much prettier if allowed to grow a little untidily, anyway. *Cistus × hybridus* (syn. *C. × corbariensis*) is naturally neat, but *C. × cyprius* likes to sprawl and should be allowed to. If its branches can lie on the ground they give the plant all the support against wind-rocking that it needs.

In the same way I like unpruned rosemaries. It is natural for them to become gnarled and woody in age; they develop character. In any case you can never rejuvenate them by cutting back. Nor can you lavender. Start again from rooted cuttings. The most lavender will take is clipping back of its previous year's shoots.

I wish I knew more about the *Hamamelidaceae*: parrotia, corylopsis and sycopsis, for instance. I'm told that fothergilla responds excellently to cutting back, and from the appearance of my own *F. monticola,* which branches freely from the base, I can well believe this. On the other hand *Hamamelis mollis,* the Chinese witch hazel, is generally unresponsive. I cut large branches for the house, but the bushes are rarely stimulated into fresh growth anywhere near where the cuts were made. But they do, from time to time, make good with young shoots produced low on the bush (beware confusing these genuine shoots with rootstock suckers from ground level).

Rhodendrons are generally much too close-planted by enthusiastic owners lacking experience and acres. Space and labour shortages will also prevent subsequent thinning and transplanting, which would otherwise be the most satisfactory solution. Cutting back therefore has frequently to be resorted to and it works excellently in most cases, young growth breaking from thick, rough old trunks. The exceptions are those species whose smooth, peeling trunks are among their principal attractions: *Rhododendron barbatum* and *R. thomsonii,* for instance. Bean gives us the reason (*Trees and Shrubs Hardy in the British Isles,* Vol. III, p. 551): in sloughing off the outer bark they also discard their dormant growth buds, we are told. It seems a strange arrangement, but we don't have to believe it. The answer, anyway, is that you should not take liberties with any smoothy.

It has always struck me as sad that the wives of rhododendron barons are never allowed loose among the rhodies that they may help themselves to armfuls of flowers and foliage, for these look marvellous in flower arrangements. Fortunately I can do as I please in this respect. I have been interested in the results of cutting into three-year-old wood of *R. fortunei* subsp. *discolor,* for this purpose. My particular clone flowers rather early in June, and its long straight stems are admirable cutting material. The same stem length makes this a leggy species, so you are actually doing the shrub good and encouraging it to bush out by your plundering tactics. From just below each cut I made, a ring of up to half a dozen young shoots quickly developed. They only grew a couple of inches that season, but forged ahead and made terminal flower buds in the next.

Camellias, so often associated with rhododendrons, respond to cutting back with the same alacrity. 'Cherry' Ingram had a *Camellia* × *williamsii* cultivar (it may have been 'Donation') that had been drawn so high by an overshadowing tree that it flowered out of sight. The tree went and Cherry cut his camellia back to 3-ft stumps without a vestige of greenery anywhere. They broke freely right up to the scars and without any dying back whatsoever. It was an amazing sight.

One is so inclined to take it for granted that hebes (the shrubby veronicas) can be cut back as hard as you like with impunity that it comes as a surprise to discover that this is not invariably true. I have two *Hebe* 'Mrs Winder', which are grown as foliage plants since they flower little but their narrow, elliptical leaves have a purplish cast which is especially pronounced in winter. Indeed, I have just been out to study the form on a frosty January morning and the better of them does look perfectly delightful, its darkness lightened with a margin of white rime that outlines each leaf distinctly as an individual.

These two bushes, growing side by side, became so large that they had killed out too large a slice of adjacent turf. So I cut them hard back in the early months of 1975. I don't know if the year is significant or not, but certain it is that a serious drought developed (not as bad as in 1976, admittedly) and that whereas one of these hebes forged ahead and was again, after two years, a perfect dome of young shoots rising to 3½ ft, the other looked as though it was a goner. Eventually it sprouted two weak shoots from near ground level and I had to saw away its main stump. I should do better to replace it – or extract and not replace it.

Most hebes grow so fast and satisfactorily from young rooted cuttings that replacement of the old chaps with a youngster, rooted a year or two in advance of the exchange, is the most sensible course of action. This applies especially to showy cultivars like 'Gauntlettii' and 'Great Orme' with a long flowering season. On an old specimen this season is much reduced and so are the length and quality of its flower spikes. Whipcord hebes – those with small scale-leaves bearing a notable resemblance to dwarf conifers – are not quite so responsive to cutting back as the other kinds, though it can work on them, too.

Turning now to lilacs, on which I have accumulated quite a bit of experience. They all break very well from old wood and so do the closely related privets, but they are liable to a number of fungal diseases and if you're unlucky severe treatment may be fatal. It is certainly wise to spread any cutting-back operations over two or more seasons rather than slaughter an entire bush in one glorious encounter. I tell you that without having the faintest intention of following the same advice when I feel otherwise inclined.

Any of the numerous cultivars of the common lilac, *Syringa vulgaris,* will become overgrown in course of time. It may be enough simply to thin out all the weak old twigs and branches in the centre of a bush, leaving its outline virtually untouched. By admitting light to the previously darkened centre in this way, new growth buds will be stimulated into action and strong young shoots will develop.

Or you may go a step further. As well as the above-described thinning out, which is a sound practice every five or six years in every lilac, you can remove one or two really large branches, right back to their point of origin low on the shrub or even at ground level, for many lilacs spring as multi-stemmed specimens from the ground or near it. I sawed a branch of this kind from my double white 'Mme Lemoine' three seasons ago, making the cut horizontal and as close to the ground as I could – actually about an inch above. From this cut it has thrown four young shoots which are now 6½ ft high. They would have grown taller still in normal seasons, but two of the three were abnormally dry.

In the last resort, when a bush has thoroughly disgusted and inflamed you, it can be cut back to stumps all over, all in one operation (simpler than grubbing it and almost as satisfying a method of letting off steam). I did this to two bushes that were growing side by side, both pink-flowered cultivars, one called *S. × josiflexa* 'Bellicent', the other *S. × swegiflexa* 'Fountain'. They were some 15 ft high and nearly as much across, dense and shapeless with blossom on the perimeter, not very attractively displayed. I sawed them back to 3½ ft one winter. In the next season they sprouted vigorously, making 3 ft of young growth. But 'Bellicent' faltered towards the end of the growing season, wilted and died. 'Fountain' forged ahead and flowered in the second spring, some sixteen months after its treatment. It is now, five years following the operation, 10 ft tall.

The growth of a wintersweet, *Chimonanthus praecox,* is not unlike a lilac's. This is just as well because it is particularly prone to attack by the coral spot fungus, *Nectria cinnabarina,* which is easily recognized by the rash of red pustules growing over the dead wood from which it is drawing nourishment. Whole branches of wintersweet can be killed by this fungus, but the entire shrub will never succumb because it can always throw new young shoots from the base.

Trading on this fact, I once cut the oldest of my wintersweets right to the ground except for a few wispy shoots. It had become a tired bush in the course of forty years, and had largely ceased flowering on branches that had themselves almost ceased to make new growth. I hoped, by my treatment, to rejuvenate it and I did. It has never looked back.

Many shrubs need regular pruning by the annual removal of old, flowered branches, making room for the young unflowered shoots. Such are philadelphus, weigela, deutzia, dipelta, kolkwitzia, most ribes and spiraea. But what if one of these has never received this pruning? Your wise and cautious approach will be to sort out and remove a few old branches each year, so that the entire shrub is replaced after four or five years. Alternatively cut the whole thing to the ground and have done. Such treatment will inevitably reduce its vigour in the short term, but it will throw new shoots from the base all right and its health will presently be completely restored.

Stooling is often an effective practice with ornamentals. When performed annually it perhaps hardly comes within the scope of this already long chapter. Nevertheless, I shall mention its value on *Cotinus coggygria* (*Rhus cotinus*) in any of its purple-leaved cultivars. The unpruned specimen will grow to 10 ft, perhaps, and will flower for you, which is nice, but if it is the smoky results of flowering that you want, take my tip and grow straightforward, green-leaved *Cotinus coggygria* itself. The foliage on an unpruned specimen becomes progressively smaller and dowdier. Even at this stage you can transform it by cutting to the ground. Regularly stooled, your purple cotinus will make wand-like shoots to 3 or 4 ft only, bearing the largest leaves of wonderful vitality, for new young leaves will continually unfold on its extension growth right through the summer. Imagine this with the glaucous foliage of *Melianthus major* or of *Berberis temolaica,* also cut back frequently.

Stooling also gives you the largest and most glamorous leaves on the pinnate-leaved sumach, *Rhus glabra* 'Laciniata', on *Ailanthus altissima* (tree of heaven) and the great hearts of *Paulownia tomentosa.* Also on the golden catalpa, the purple filbert (*Corylus maxima* 'Purpurea'), and on all those elders (*Sambucus*) that are grown for their foliage rather than their flowers. I have two plants of the cut-leaved golden elder, *Sambucus racemosa* 'Plumosa Aurea', in my Long Border, which I prune annually like a *Buddleja davidii* by shortening all their last season's shoots back to within a pair of buds from their base. Over the years, a woody foundation of stumps has built up, eventually 5 ft high. When the coppery-tinted young foliage expands in spring, I am uncomfortably aware of these ungainly stumps until well on in the growing season, and comparison with photographs of my plants when they were youngsters brought home to me the fact that they looked far prettier then, with young foliage down to the ground and no stumps to mar it. So, leaving one of my plants as usual in case of accidents, I last winter cut the whole of the other off flush with the ground. Its response was miraculous, and even in the driest growing season ever recorded it made 7 ft of growth. So I am rid of my stumps and have an utterly rejuvenated plant.

There are some willows in some positions that you regularly pollard. *Salix udensis* 'Sekka', for instance, with its intriguing flattened and twisted stems. Assuming that you enjoy this aberration, it is much more pronounced on the young growth following a hard pruning than on unpruned specimens. Then again, the red stems of *S. alba* var. *vitellina* 'Britzensis' (syn. 'Chermesina') and 'Chrysostela' are brightest and most abundant on the young wands that follow annual pollarding. And the young purple stems of the violet willow, *S. daphnoides,* need encouragement in the same way. When this has been going on for a number of years, a great bulbous mass of stub ends builds up at the top of your willow's trunk or trunks, and you should make a clean sweep of all this accumulated evidence by lopping off the entire head. Pollarded willows need not be accepted as permanent fixtures. Eventually you will want to replace them with young rooted cuttings and start all over again. It should, in any case, be noted that butchered willows are liable to be attacked and killed by the silver leaf fungus, *Chondrostereum purpureum.*

Fatsia japonica is wholly cooperative. It has a slight tendency to sucker close to its trunk and from below ground level, even when unpruned, and so however barbaric you choose to be it will come back smiling. Fatsias have rather a way of growing larger than you thought they could. But before you take action, do ask yourself whether action is really called for. A giant specimen is a splendid feature. If you want room for other plants, cut the fatsia's lowest branches away and you can then plant right up to it with ferns, hostas, hardy orchids, snowdrops and other shade-tolerant things.

All the viburnums I can think of respond equably to harsh treatment. *Viburnum opulus* 'Compactum' is one of my favourite mixed border ingredients, for its crimson berries are showy for three months from August and it flowers prettily if briefly in May. But it had got up to 7 ft tall and was still rising, so I have tackled it in two stages. I cut half its stems on different parts of the bush right down two winters ago. From these cuts it has thrown a good quiverful of strong young shoots which will flower next spring, so I can now remove the remaining branches which have till now fulfilled a caretaker rôle by cropping for me until the young shoots were ready to take over.

I once had occasion (described in *The Well-Tempered Garden*) to reduce a strawberry tree, *Arbutus unedo,* to a skeleton in November and this broke into new shoots all over its rugged trunks the following May. I have never needed to do this to a pittosporum, but from the way my *Pittosporum tenuifolium,* which I try to grow as a tree with a clean trunk, continually breaks into young growth from old wood right down to the base, I think it pretty certain that I could have my way with it.

And you can cut freely into the oldest wood on any escallonia, which is just as well when a hedge of it has got out of hand. The *Escallonia* 'Iveyi' in my Long Border is gigantic. It is not one of the hardiest of its tribe, and anyway I don't want to lose its blossom if I needn't, so my method with this is to remove a proportion of its oldest branches every year in late winter, replacing and reducing it in many imperceptible stages.

Where *E.* 'Apple Blossom' was bulging over a path, I wanted to hack into its old wood all in one go. I waited till it had flowered and did it then, in early July, rather hoping that it would grow enough in the tail end of that season to flower for me the very next summer. That was being greedy. It made some growth but not much. Lots the next year, of course, on which it flowered the year after. I once cut my 12-ft *E.* × *langleylensis* back to 3-ft stumps, and that responded almost too well. I finally got rid of it.

In doing so I had to cut a mature *Clematis* 'Huldine', which was growing lustily all over it, hard back into five stems ½ in. and ¾ in. across. They all broke into new growth straight away; I have just been out (12 January) to check on them and they have large young buds breaking from the lowest joints on the thickest wood.

But clematis differ as between varieties and species in their response to cutting back. If you cut a very old *C. montana* down into a rope-like trunk, you may set it back so seriously (even without actually killing it) that you'd have done better to replace it with a young and uninhibited plant. I have killed a *C. macropetala* aged not more than ten years by hard pruning, and *C.* 'Columbine' didn't like me at all when I cut it all back. After two years I still don't know whether it will fully recover. If you had a 'Perle d'Azur' or a

'Comtesse de Bouchaud' with a single, thick trunk some eight or ten years old, which is the way they're apt to grow, I should expect it to die if cut hard back to 2 or 3 ft. But if it was in the habit of making shoots low down then you could do what you liked with its main trunk, as I have discovered with *C.* 'Jackmanii Superba'. The more stems a clematis has the thinner they'll be and the more likely to respond well to cutting back. 'Nelly Moser' is good in this respect.

Climbing honeysuckles (*Lonicera*) break well from old wood, and so, to the best of my knowledge, do most of the bush types such as *L.* × *purpusii* and *L. fragrantissima*.

We have a very ancient *Skimmia japonica* (sixty years old, at a guess), which was not merely rather a nuisance projecting over a path but some of whose branches kept dying back in an indeterminate way. A none too healthy specimen of this age is unpromising material for rejuvenation, but (having, as usual, insured myself with a young replacement) I was in a kill-or-cure mood. Having cut the skimmia to 1½- and 2-ft stumps in spring, a long time elapsed before anything happened. Eventually it broke, mainly low down, and quite a lot of the stumps died. It has been making a gradual and reasonable recovery over five or six years. It is a male, and when I had first cut it back I had to stand a jam-jar of flowering male twigs, gathered from another part of the garden, in its place to act as pollinators to the adjacent wives.

The closely related Mexican orange, *Choisya ternata*, can always be restored to pristine youth. Large branches are easily broken under the weight of snow. Cut them out and within a year or two you'll have forgotten all about the accident. You may sometimes have to do duty for snow: a choisya consisting entirely of old branches tends to lose vitality. Leaves and flowers become progressively smaller and the foliage takes on a yellowish cast – especially in very sunny positions. If you cut a branch or two right out you'll be amazed at the transformation in the young growth thus stimulated.

Phlomis fruticosa, the Jerusalem sage, is another martyr to snow damage but again it breaks well, however old, and so does *P. chrysophylla*.

If only gardeners would take saw and secateurs to their mahonias more readily and regularly. As a group they are first rate both for foliage and flowers, but as shrubs they are ruined by the let-alone treatment for they become unbearably leggy and scrawny (although in great age this condition develops its own appeal). *Mahonia* × *media* 'Charity' is the worst case in point. It is wonderful to have such a showy shrub in December, but what an eyesore as you see it nine times out of ten, with that upright swept look, a mass of naked stems topped by tufts of leaf and flower. Both its parents, *M. lomariifolia* and *M. japonica* and all the many hybrids derived from this cross: 'Buckland', 'Lionel Fortescue', 'Winter Sun' *et al.* need going over every spring, shortening all those stems on which the wood is showing. They'll break freely from your cuts and flower from the young rosettes the next autumn, winter and spring.

Mahonia × *wagneri* 'Undulata' (a *M. aquifolium* cross) is one of the handsomest evergreen features in my Long Border, especially in midwinter, when its glossy foliage is rich purple, but it would grow into a very large and space-consuming bush did I not bite deep into its framework over the whole specimen every five years, immediately following its April flowering.

The vanilla-scented, late-winter-flowering *Azara microphylla* habitually grows larger than the available space permits. I have not been lucky in my operations on this. I cut our original plant back to its trunk, leaving about 8 ft of this, some thirty years ago now. It broke well, but then went into reverse like other shrubs I have described, and died. Its replacement now needs treatment, for it has become unacceptably scraggy. I shortened some of its worst branches two years ago, and the stubs have remained absolutely bare. On the other hand, a friend in Dorset with the same problem cut the whole of his azara back to about 5 ft and it responded splendidly with never a backward glance.

Shrubby members of the *Asteraceae* are pretty good subjects for rejuvenation. The popular grey-leaved *Senecio greyi* or *S. laxifolius,* now designated *Brachyglottis* 'Sunshine' (it is a hybrid), becomes hopelessly messy if not brought strictly to heel every two or three years. Those who dislike its flowers cut it back every year. I like them, and only prune it heavily after a heavy flowering. *Brachyglottis rotundifolia* (syn. *Senecio reinholdii*), with its thick leathery leaves and undulating margins, is one of the most beautiful of all foliage plants that we can grow outdoors and although its hardiness is by no means to be relied upon, except in coastal gardens, it is very much hardier than is generally credited. I cut mine hard back one spring, and it was by no means pleased. It took the whole summer to recover its looks again, and then most of the young growth came from low down, not from the cut stems.

Olearias, the daisy bushes, all respond excellently to cutting back and break from the oldest, thickest wood. *Olearia macrodonta,* used for shelter belts and hedging near the sea, can be restored to order in the same way as escallonias, when the need arises. *Helichrysum splendidum* is the only hardy shrub member of its genus and this can be cut back into bare wood – a good thing to do every four or five years. But if you cut it into a really old trunk near the base, as I once did, you'll kill it. The nearly related *Ozothamnus* respond well to surgical treatment, as I have lately discovered on an *O. rosmarinifolius* which was 8 ft high and is now 2½ ft and sprouting all over.

Finally the rose family. These are good subjects on the whole except, as I said at the beginning, where the disease factor comes in. You can cut the cherry laurel and Portugal laurel hard, and they often are so massacred, but the latter, in particular, is subject to silver leaf, so you shouldn't omit your wound dressings. These belong to the genus *Prunus*, of which plums, cherries and peaches are the trickiest members. Still, there are times when you have to risk all.

I had a thirty-year-old standard of the winter cherry, *Prunus × subhirtella* 'Autumnalis Rosea', whose appearance had been ruined by repeated bud-pecking by bullfinches. It was a mass of long, bare writhing branches; flowers and leaves were carried only at the extremities, where the solidly built bullfinch was unable to perch. In the spring of 1969 (3 April, to be exact) I sawed all the branches back to within 1, 2, 3, 4 and 5 ft of the main crutch. I pared the rim of each wound with a sharp knife so as to facilitate callusing and, of course, I applied a wound dressing. The tree never faltered and there was very little dying back at stub ends. Its strongest young branches grew a good 5 ft in two seasons. There were hardly any flowers in the first winter but masses in the next. The bullfinches are delighted.

You can do anything you like with *Chaenomeles speciosa* cultivars, the Japanese quinces, but the harder you cut them the more they'll be induced to sucker, which can be a nuisance, and this should be borne in mind with a number of other shrubs I have mentioned. It is a particular nuisance when they are grafted plants and the suckers are of the stock: rose, rhododendron, lilac, witch hazel, etc. This reaction is only to be expected, when you come to think of it. You remove the branches from a shrub with a strong root system. Its energies have to transfer somewhere and if the roots or rootstock can make growth buds, your cutting back will stimulate them to do so. Another reason for not cutting Japanese quinces hard if you can help it is that they will take several years of vigorous growing before they will settle down to flowering again.

Kerrias, single and double (*Kerria japonica*) are seldom pruned regularly, though they should be. You can take the bull by the horns and cut them right to the ground after flowering, in spring. I did this once and had little blossom the next year but was back to normal after that. The best way with shrubby potentillas is to let them alone for most of the time, in which case they start flowering in May. But every five years or so, cut them hard back. Their response is perfect but the onset of flowering is delayed till July.

Cotoneasters respond to any amount of bashing. If I want to get rid of a long-established self-sown *Cotoneaster horizontalis,* I cut it down as low as I can. A forest of young shoots duly appears and these are treated while tender and green with a hormone brushwood killer (unfortunately this is largely ineffective against *Chaenomeles* suckers).

You often see a bulbous old pyracantha that is clipped annually to prevent it pulling the house down but is thus effectively prevented from flowering and fruiting to any worthwhile extent. If you cut it back to its trunk so that it will not need pruning again for several years, it will be restored to productivity – for a time. But pyracanthas are really only worth harbouring where they have room to grow freely without interference.

And so to roses. Like clematis, they are variable in their response according to type. If you cut a thicket of old Rugosas to the ground they respond beautifully, making a 3-ft dome of young shoots in the next summer. Flowering will start six or eight weeks later than it otherwise would – in July instead of May. One must add, however, that true Rugosas (not Rugosa hybrids) are in question here: such are 'Fru Dagmar Hastrup', 'Alba', 'Scabrosa', 'Roseraie de l'Haÿ' and 'Blanche Double de Coubert'. Furthermore, they should be on their own roots so that all the suckers they freely produce are their own.

The stronger Gallicas, like 'Tuscany' and 'Belle de Crécy', can be cut back all over to 3 ft immediately after flowering (they only flower once); the dwarfer *Rosa gallica* and its stripy sport 'Versicolor', back to 2 ft.

Powerful repeat-flowering shrub roses like 'The Queen Elizabeth' and 'Chinatown' need to be taken back to 3 ft every winter if they are not to exceed 7 ft when flowering.

If you have an old specimen of a climbing rose that has long since ceased to make a productive young shoot from the base, you must try to cajole it into doing so before taking drastic action. Feed and water it heavily, and reduce it at the top by, say, one third. Once you have a strong basal shoot wherewith to renew your plant, the rest of it can soon be eliminated.

PRUNING THE HYDRANGEAS

It is as important to delay your hydrangea pruning till spring as it is unimportant when between autumn and spring you prune your roses. The shoots that are going to produce the hydrangea display need first to overwinter (and overspring, for that matter) without being destroyed by frost, and the fact that they are surrounded by other branches, which you will eventually prune away, and are surmounted by last year's crop of dead blossoms, all helps to see them safely through.

I enjoy any excuse for contact with hydrangeas and this pruning is our first get-together of the year. As I stagger with armfuls of their branches to the rubbish heap, I hug them closely, as much to extract the sharp satisfying scent from their expanding foliage as to prevent dropping a trail of them on my way. I have about fifty bushes, a mental reckoning tells me, but scattered as they are through different parts of the garden, this by no means makes a numerous impression, and there are many more that I should like to find room for.

Most hydrangeas grow to a pretty fair size, but regular pruning keeps the majority of the best-known *macrophylla* types to 3 or 4 ft high and wide. Combined with generous feeding (again in spring) it ensures that the bushes remain healthy and productive of blossom over a long season. By constantly removing the older, weaker branches you are continually making room for and encouraging the production of strong new shoots, and these, from the centre of the bush, should on average grow 3 ft in a season.

Never wear a jacket with loose cuffs and tails when out hydrangea pruning, as you are sure inadvertently to catch the most promising young buds in them and they'll snap off all too easily; a snugly fitting jersey is ideal. You'll need a slim saw as well as secateurs. Maybe the branches you wish to remove are none of them too thick for secateurs to manage, but a saw is sometimes handier for getting into a tight corner low in the bush. And if you're removing a branch completely, you do want to cut it as low as you possibly can, otherwise, looking into the future, there'll be increasing difficulty in taking branches right out if they're surrounded by thick old stumps.

'Where on earth do I start?' you wonder, as you contemplate the redoubtable tangle of an unpruned bush. If you make a regular practice of pruning your hydrangeas each spring, there's little difficulty. You note the branches that are covered with old flower heads. They have a gnarled appearance, with pale bark. Cut them out, taking them back to the first strong unflowered young shoot or, if there are none such, as is often the case, back to the ground. By the time you have gone over the whole bush in this way you'll find that you've automatically dead-headed it also. All that will be left will be a forest of strong, unflowered young shoots.

If your bushes have not been pruned regularly, remove half to two-thirds of their oldest wood. This will leave some fairly hefty gaps, but don't let that make you nervous. Gaps are good; they admit light to the centre of the bush and will soon be filled by the strong, productive young shoots that you wish to encourage. Next year you'll finish the job by taking out all the remaining old branches.

Hydrangeas that are never pruned become enormous and lopsided and their owners may, in desperation, cut entire bushes right to the ground. Done in spring, this really isn't at all stupid. A year's flowering will be lost (with the exception of a few enormous terminal inflorescences on the young shoots at the very end of the season, if October remains congenial), but the bush will have been rejuvenated at a stroke and will be set, if its shoots survive the winter, to give the display of its life in the following year.

If these young shoots are killed, and then again killed or crippled in succeeding winters, have the plant out and either remove it to a more sheltered situation or throw it away. There is considerable variation in hardiness between one hydrangea cultivar and another, and there is also considerable micro-climatic variation within different parts of your garden. You have to learn to fit the one to the other.

A few notably hardy, white-flowered hydrangeas, of which *Hydrangea arborescens* and *H. paniculata* are the best-known species, make a practice of flowering at the tips of their young, current season's shoots. All growth can therefore be shortened hard back each spring, as you would the common butterfly bush, *Buddleja davidii*.

How hard you prune them depends partly on how high you want your bushes to grow, but also on how large you like the inflorescences to be. Very hard pruning results in a few very large flower heads. In the case of *H. paniculata* 'Grandiflora', wherein all the florets are sterile and densely packed, this can look more like a weapon to snatch at and hit a recalcitrant relation over the head with than a gentle contribution to the flower garden. But in *H.p.* 'Floribunda' and the very similar 'Tardiva' (they both flower from August to October), the large sterile florets are infilled with a foam of tiny fertile blossoms, and even the largest inflorescence will never look out of scale. You should certainly never leave this type of hydrangea unpruned, or there will be too great a loss in quality and size of inflorescence. There is quite a case for pruning the weak-stemmed *H. arborescens* to the ground every winter.

Hydrangea paniculata 'Praecox' is exceptional within this species in flowering from the old wood (seldom from the new) and therefore much earlier than the others. It should be allowed to make a 6- or 8-ft bush, removing old branches only as they become weak and unproductive.

MAINTAINING MATURE HEDGING

Formal hedges to a garden are as walls to a house (or garden, for that matter), its basic framework, but they need considerably more upkeep. Much has been written about the choice and establishment of new hedging, but I want to say a bit about the problems and obligations imposed upon us by the maintenance in good condition of the mature garden hedge. I shall take yew as my example, because it is the handsomest hedging material available and because there are many hundreds of yards of it in this garden at Great Dixter. Most of the points that apply to yew apply also to other forms, while I have already discussed their individual reactions to being cut back.

A hedge is not as easily kept healthy as a tree, because every time you clip it you are obliged to remove all its youngest and strongest shoots and leaves (lawn turf provides an exactly similar parallel). This is inevitably weakening. So is the close spacing of the plants within a hedge. Each is competing with its neighbour, its growth being entirely halted at the interface. The moral, here, is not to plant too closely.

In any case feeding is *de rigueur,* and a slow-acting organic manure is the most satisfactory. We give a surface dressing of blood, fish and bone compound every year around February, but before applying this we clean out the hedge bottoms. Competition from foreign bodies for light and nutrients has to be prevented. One of the worst weeds to take hold in a hedge (and it always will) is ivy. It must be pulled out, as must self-sown saplings of ash, sycamore *et al.* Then, of course, grass and other herbs will grow into the hedge bottom, so you must allow a clean, cultivated gap between this and the nearest other plants. Even if this is merely a lawn, six clear inches should be allowed. Not more, however; use of the half-moon edger is fatal.

Summer competitors which can only be dealt with in summer because, being herbaceous, they are only visible then, are bindweed and the similarly growing black bryony (*Tamus communis*), which carries handsome swags of red berries on female plants in the autumn. Its root is a large tuber. Creeping sow thistle and the creeping field thistle will also find an entry. Their seeds blow in from neighbouring farmland and get hung up on the hedge face, thence eventually falling to the ground. All these can be tackled with selective weedkillers.

If you keep on removing weeds like grass from the bottom of a hedge, the soil will gradually disappear with them. On a slope, erosion will have the same result, and the yew roots will be exposed. It will then be necessary from time to time to import topsoil (and more weed seeds), this being most easily obtained, as a rule, from the accumulated unburnt remains of a bonfire site.

Growth in parts of a formal hedge may, in course of time, come to a virtual standstill. This is a danger signal and may result in death if nothing is done about it. The cause is usually bad drainage. Another signal is the growth of grey lichens on the outer hedge shoots. It is not the lichens that need controlling but the conditions that produced them. You'll not see lichens on a hedge that is growing freely.

You must ensure that drainage is good at the time of planting a hedge. One old method on heavy soils like ours was to plant the hedge on an artificial ridge. My father laid tile drains under all our hedges before planting them, but they were, at 2½ in., of insufficient diameter. But even large-bore land drains may be expected to get blocked in time, or there may be sinkage somewhere along the line that prevents an even downward flow. If the hedge is at the edge of an old lawn, you may get considerable compaction over the years and such an accumulation of fibre mat in the turf that water cannot get away.

To retain its health, a hedge's greatest thickness should always be at ground level, tapering upwards. The slope thus produced, called in building terms the batter, allows light to reach the hedge's bottom branches so that they grow as healthily as those at the top. If you start with a vertical-sided hedge, the bottom is shaded and the top grows more vigorously at its expense, thus aggravating the situation. You soon get that familiar feature in the mature hedge of an ugly, unbalanced-looking overhang. As my father wrote in his book on yew and box, published in 1925, 'yew hedges should have from 2 to 4 ins. batter to each foot of height'. Apart from the light factor, this makes the hedge look substantial.

If your hedge has developed an overhang or if, as is eventually inevitable even when plenty of space was allowed originally, it encroaches on both sides of a path going through it so that two people can no longer pass arm in arm nor a capacious wheelbarrow solo, drastic action must be taken. So resilient is yew that you can cut it hard back into old wood, even back into a hedge unit's trunk, and it will break into fresh, green growth. This is a frightening experience to the uninitiated, as the immediate effect of cutting back is hideous, but it always works with yew. Do any cutting back you must in late autumn or winter.

Another occasion when it becomes necessary, with yew, is when a branch in a hedge or topiary specimen suddenly dies. A disease is the cause, but it fortunately acts in a sporadic manner and never kills an entire bush, only a branch here and another there. It draws attention to itself by turning the dead yew leaves vivid brown, and the affected limb must be cut right out. This leaves a nasty hole, but it is surprising how quickly it will fill in.

TOP-DRESSINGS AND ROOT PRUNING

The art of growing bonsai has developed its own mystique, but when you come to study the subject it turns out that common sense and patience are the requisites for success, as in other gardening departments. Root pruning is commonly supposed to be one of the secrets of successful bonsai cultivation, and the very thought of it strikes terror into the hearts of would-be initiates. How is it to be done and when? they wonder, quite forgetting to ask why it should be done at all, for the docile beginner is anxious to follow the rules without questioning them.

I do not grow bonsai myself, but I have a friend who is expert in this craft and I can confidently quote his explanation. Some of the soil in a bonsai's root ball needs replacing from time to time as it becomes exhausted of nutrients. You therefore turn it out of its container, perhaps once in three years, cut away the outer layer of peripheral roots and remove the soil in this layer, thus making room for fresh soil into which the tree can push fresh roots. It is a rejuvenating process, in fact, and is not performed with the object of dwarfing the tree, as is often supposed. The tree's size will be controlled by the size of its container and by your pruning of its shoots.

This kind of root pruning is really nothing more than top-dressing in reverse. The only tricky part about it that I can see is making sure that you do in fact fill all this newly won space around the inside of the container, with fresh soil and without leaving gaps and air spaces.

The normal way of rejuvenating a pot plant, without drastically washing or shaking the soil away from its roots, is by removing the top layer of soil with the tip of a widger or of a pointed wooden label, prodding about until you have extracted as much as you dare without seriously damaging the plant. Then you replace with fresh compost – a strong mixture such as John Innes No. 3.

Clearly this could not be done with a bonsai because part of your picture, in this case, is created by the mounded soil surface, overlaid with mosses, out of which rise some of the tree's own gnarled roots. All these stage props must be preserved.

The top-dressing of pot plants which you do not actually want or need to move into larger pots is something that should be done at least once a year. It can be carried out at any season. A convenient time is often just after a plant has finished flowering. Thus, *Coronilla valentina* subsp. *glauca* and *Genista canariensis* (the genista of florists) are scented shrubs with yellow pea flowers whose season is at its height in April. When it is waning, cut the plants back by about a third all over, top-dress them, and stand them in a suitably sunny position out of doors.

With bulbs like hippeastrums and vallotas, it is normal to top-dress at the start of their growing season, which would be January or February in the former case, April in the latter. Another moment when I do a lot of top-dressing is when I'm emptying a cold frame or my greenhouse. Unless you make a point of shifting your pot plants in this way every year, their health will suffer and you will get a nasty accumulation of liverworts and weeds in and all around them.

Root pruning is not normally practised on pot plants apart from the special case of bonsai. It can be practised on individual trees in the garden or orchard which are making excessive growth without settling down to the business of flowering and fruiting. You should, however, never need to root prune in the whole of your gardening life. Fruit trees, for instance, are grafted on dwarfing rootstocks to make them productive at an early age and while still small. This is a far more efficient way of getting the desired response than by root pruning, which is really a measure of despair. Far better, in most cases, to rid yourself of the unproductive tree that won't stop growing.

If you plant a Brunswick fig and hope for a luscious harvest you will have to wait many years for it, while allowing the tree to grow enormous. Branch pruning will merely result in masses of soft, unproductive wood. You can restrict the tree's roots by growing it in a container and this, indeed, is an excellent way to grow figs and have them fruiting young, especially under glass with a view to obtaining two crops a year instead of the one crop which is all we can manage outside in our climate. But even then you should choose a precocious variety like 'Brown Turkey' or 'Black Ischia'. I must add that nearly all the figs I see grown in this way by amateur fig-fanciers are starved and incapable of making decent growth, let alone fruits. It is essential to water and feed a container-grown plant generously. Really you'll be making life far easier for yourself and for your fig if you can possibly grow it in the ground and allow it a free root run. Then it can find its own moisture without worrying you for supplies, and a precocious variety will naturally settle down to fruiting at an early age. 'Brunswick' admittedly has the largest fruits and the handsomest foliage, but it needs the largest setting to do itself justice.

There are many trees and shrubs with which you simply must be patient. No amount of root pruning or restricting, bark ringing or tourniquet applications will bring them to early maturity and productivity. The flowering dogwood *Cornus kousa* var. *chinensis*; the Chinese pagoda tree *Styphnolobium japonicum* (syn. *Sophora japonica*); *Eucryphia* × *nymansensis*; *Davidia involucrata,* the handkerchief tree; *Magnolia kobus* and many large growing rhododendrons: these are just a few examples of the plant that should be given its head without worrying yours about how to make it flower before it is ready to.

THE ART OF COMPROMISE

When we are in trouble with our plants, the experts to whom we turn for succour are apt to be perfectionist in their recommendations.

Diagnosing a nasty case of leaf and bud eelworm in some border plant you've sent in, the laboratory worker will tell you to burn the lot, including all scattered dead leaves, and fallow the ground for a year. No weeds must be allowed to take hold, since many weeds are a prey to the same pest (which is microscopic and makes your plants look diseased). Indeed, a list of known susceptibles among garden plants and weeds makes your case look pretty hopeless.

And so it proves, because a year later another but unlisted victim of the same eelworm crops up in another part of the garden some twenty yards away from the first. More clean cultivations are prescribed. It begins to look as though one will need to move gardens. Because, after all, large patches of bare earth are not pretty to look at in a flower border in the growing season.

The above example is my own case (the affected plants being *Lamium maculatum* and *Eryngium proteiflorum*), and it has seemed to me more sensible to keep on planting the affected areas until I find something eelworm-resistant, thereby retaining a garden that looks like a garden, rather than a wilderness. After all, the most pessimistic nematologist (eelworm wallah) has never claimed that all or even half the plants grown can be attacked by one given genus or strain of eelworm. I noticed that the list quoted to me did not include any monocotyledon. Perhaps an accidental omission, but so as not to be entirely empirical in my methods I planted up with hostas, hedychiums, montbretias and yuccas. After four seasons, all still seems well with them.

Rust diseases can give rise to anxious situations, some worse than others. If you get rust in your spearmint, it's not difficult or expensive to burn it and plant elsewhere with a few healthy roots from a friend's garden. Rust in roses, on the other hand, can be a more painful experience. One has heard of whole rose gardens being destroyed. Of course there are protective spraying programmes that can be brought into action, and these can simultaneously wage war on greenfly, mildew and black spot.

A professional friend who not infrequently visits the National Rose Society's trial grounds at St Albans made the revealing remark to me that he actually finds them too well kept to be a really useful guide. Few rose lovers, unless monomaniacs, are prepared to go further than half a dozen rose sprayings in a year, or to dead-head their bushes more than once a week. If fading blooms are removed daily, you never learn how frightful they can look on the bush after four, five, six or seven days (and you'll never learn this from the show bench either, of course). If every recommended spray is applied against fungal diseases, you'll never sort the sheep from the goats, as under a more haphazard but humanly probable routine.

Apropos of rose rust, there seem to be a great many different strains, some less virulent than others. My vast specimen of the double red climber,

'Guinée', was completely defoliated by rust and severely crippled for a season when about four years old, but it recovered and became more vigorous than ever.

It can be a mistake to plant new rose bushes in old rose soil. The nurseryman never dares take this risk. He must either fumigate the ground where repeat plantings are to be made (an expensive procedure) or rent pieces of land from nearby farms on a rotational system. In the garden, we are bid remove the old soil and replace with new. Not a task that will recommend itself to the average gardener. It is perfectly true that a new rose is often seriously stunted in its first season if planted on old rose soil, but this has been my practice, on a piecemeal basis, in our rose garden for as long as I can remember and I have no reason for wanting to change it. Some of the new roses I introduce die in their first season and I'm not blaming the nurseryman for that. If they survive that first year they have a good chance of becoming healthy and long-lived.

What of honey fungus, *Armillaria mellea,* with its enormous host range? Again, the only recommendations for its eradication are laborious and expensive and really they can only be applied in simple situations where dense mixed plantings are not in question.

My own worst case of armillaria was when an old apple at the top of our Long Border died of it. Not only was this tree growing out of the middle of a yew hedge, but its roots extended through a maze of shrubs and herbs. I lost many of these subsequently and some of their replacements also. Now, however, things seem to have settled down. The yew was never affected and other plants have, by and large, remained happy for the past six or eight years. I never see any armillaria toadstools: if I did I should dig out the root that was producing them. Compromise seems to have won the day.

If you cultivate plants well, they are much the more likely to grow away from the honey fungus and to show no signs of being infected by it. The fungus attacks (is chemically attracted to) roots underground, but to a large extent the healthy plant is able to make more healthy, compensatory roots than are killed by the fungus. Grow your plants well and you won't do too badly.

When a drought like that of 1976 comes along, you'll be in trouble from armillaria as well as from water shortage. Weakened by drought, many trees and shrubs were overtaken by the disease that they'd previously coped with. The primary cause of death in many 1976 cases (including the tough old *Rhododendron ponticum*) was honey fungus; only indirectly drought.

The gardener's best defence is not dramatically remedial but cautiously preventive. If you cut down a tree, always remove its stump and roots if you possibly can – and in the case of orchard trees this is always possible. The main reason for so many woodland gardens being riddled with this disease is that, in order to make room for rhododendrons and other desiderata, trees and scrub were felled but their roots remained as a source of subsequent infection. Then, again, if you keep your shrubs well mulched and regularly top-dressed with organic material, even if this is only raked-up leaves, they'll be in the best condition to combat both drought and honey fungus.

FOR AUTUMN, READ SPRING

After many years of mud-larking in my borders in late autumn, nobly attempting to beat the fast-fading light at one end of the day and a legacy of night frost showing every reluctance to depart at the other, while the threat or fulfilment of rain ensures that there is no sense of letting-up in between, I have at last come to appreciate that there is a great deal to be said for putting off the evil day.

Immoral, of course. But where is morality today? The beleaguered outpost of fuddy-duddies who mumble on about maintaining standards. Feel free is the motto here and now.

The average gardener, whom we must conceive of as being a lazy, pleasure-seeking so-and-so, has traditionally waited upon the Easter holiday before getting down to the annual tasks in the garden that are associated with the dormant season, with catching up and pressing on. For form's sake he cuts down the herbaceous stuff in the autumn, but after that it's five months of feet up while the weather does its worst.

And, do you know, I'm afraid this wretched creature, if not actually right, at least has a lot of good sense on his side. I have always preached getting as much completed in the autumn as you can, because growth moves so rapidly in spring as to saddle you with a tremendous lot to do all at once and little time to do it in. That's still true, but the fact is that if you do leave border tasks till early spring they go like lightning. Unlike the chap with prying, eyeing neighbours, I have no inhibitions about my garden looking a mess through the winter; all old top growth can be left on the plants, protecting them and the ground around them. When clearing time comes in with March, those heavy, voluminous barrow-loads that had to be laboriously cut and carted in the autumn will have been reduced to feather-light wisps. Going through the phloxes is like a game of skittles. You brush your hand against them and down they go, breaking cleanly off at a lower level than secateurs would ever have reached a few months back.

The soil between clumps and shrubs, far from being sodden and panned for lack of your having turned it over to let the frost get at it – I speak of a clay soil – is light, porous and crumbly. Earthworms have seen to that. Delighted with your let-alone policy, they have been making their channels and incorporating the leaves that were never cleared away.

Some leaves I do have to clear, even at this late date; oak foliage that has blown in is still tough and composed, but much else vanishes utterly. Mulberry leaves already unrecognizable; the weigelas, which do not shed till December, gone without trace; lilacs likewise. True, the slugs have a field day, breeding uninterruptedly in a dank matrix of gently rotting vegetation, but there is so much dead stuff for them to feast on that they don't seem to damage the live more than trivially.

Conventionally, you have to take two bites at the cherry. Autumn: cut down, weed, manure, split, plant, prune some shrubs leaving others, dig over. Spring: cultivate and destroy subsequent weed growth, top-dress with fertilizer, complete pruning.

But if you leave everything or almost everything till March, you can do the whole boiling at a stroke. The pruning you would have done in autumn – weigelas, philadelphus, lilacs (when they need it) and such-like – can happily be deferred. On the other hand you may well take the view, especially if you're an old hand at the game, that somewhat tender shrubs or sub-shrubs, normally left till April, can be shorn back a month earlier. I am thinking of santolinas, *Senecio cineraria,* fuchsias, penstemons. The rather hardier buddleias, indigoferas, tree mallows and *Lonicera nitida* 'Baggesen's Gold' are no worry. Hydrangeas can be thinned and dead-headed. Roses, in whose liberal mixed-border use I am a strong believer, can be pruned at any time between November and April. So they'll fit into a late overhaul.

Splitting and replanting of perennials can go forward as usual. Is it too late to divide some things that I have always instructed should be moved in the autumn: aconitums, veratrums, doronicums? Take no notice of me (I never do). They might be checked rather more than is necessary, but most plants are tough. (They need to be.)

There are other advantages. You will be able to catch many young seedlings of annual weeds that germinate in spring: the ivy-leaved speedwell, *Veronica hederifolia;* cow parsley; goose grass; not to mention our old enemy the hairy bitter cress, *Cardamine hirsuta,* otherwise known as jumping Jesus. Its seeds are flung straight into your eyes as you stoop to remove a ripe plant.

You can now see, in order to dig out, celandines and wild arums, should they be troublesome. By the same token you can also see bulbs well through, especially tulips, and thus not damage them inadvertently, as can so easily happen during autumn operations. You may even be inspired to split up and replant large old tulip clumps, though they are already growing strongly. This move will ruin their current season's display but does not have to be rejected on that account. Ideally you would wait till flowering and the season's growth were completed, say in late June; lift then and replant in the autumn. But by the end of June the border will be yards high and brim full (let's hope). You'll never be thinking of tulips then or find them if you do remember. One must be flexible and do things when in the mood.

I am not, in fact, suggesting that border work postponed till March should in any way be skimped. (Morality comes creeping back.) That is unnecessary and unsatisfactory and will lead to a gradual deterioration from year to year as bindweed, ground elder and other perennial invaders build up into massive, throttling colonies until the situation becomes intolerable. A holocaust is then pronounced, the entire contents of the border are dug out and set on one side, and a heroic attempt is made to excavate the perennial weeds. The plants (or some of them) are then returned, but so are some overlooked weed roots and off we go again on the next downhill run. All very depressing.

A border of mainly perennial contents, whether herbaceous plants, shrubs, bulbs or (best of all) a mixture of every kind of ingredient – such a border should continuously improve for as long as it is yours. To make this possible you only need to give it your whole attention just once a year. Scrutinize each plant or group of plants as you come to it and see to its fundamental needs. No action will be called for in a number of cases, but each case will have received a hearing. March is probably the ideal moment for killing the largest flock of all those proverbial birds with one teeny weeny stone. The rest will be a matter of routine maintenance and the border, especially if watered in times of drought, will wax in strength and beauty.

YOUTH AT THE HELM

Young plants perform, on the whole, very much better than old. I have already said something on this subject as it applies to shrubs and their rejuvenation when I wrote on cutting back. Here I shall concentrate on herbaceous perennials.

So often we take it for granted that a plant, once established, can be left to get on with it indefinitely. Occasional encouragement with mulches or fertilizer is the most that will be required of our energies. Horticultural writers and broadcasters are largely to blame for this philosophy. They want to make gardening sound easy so as to keep as large a band of disciples as possible in a complaisant state of mind. Perhaps it is these gardeners themselves who are really the trouble. A people gets the government it deserves, so they say, and the saying has a wider application. If you don't tell people what they want to hear you're likely to lose your audience.

The number of plants to which the epithet 'ground cover' (with all its undertones of result without effort) has been applied is quite extraordinary. Nearly all of them will repay not merely initial but repeated effort. I had thought that *Geranium macrorrhizum* was an exception to my rule, and that it really would continue to give of its best year after year by simply being ignored. Not so, for having replanted my old colony in early summer, after its spring flowering was completed, it became another creature. Such was its renewed vitality that in late summer it flowered all over again.

One aspect of young plantings in well-prepared ground is that they often flower out of season, which may or may not be an advantage, but they also flower for a much extended season and with better quality blooms. You will notice this in a nursery, where stock is split and lined out anew every spring (often not till May, be it observed; the amateur needn't feel he's too late on more than half the occasions when he does so feel); the flowers you see in summer and autumn are finer than anything you're accustomed to in your own garden.

I will give some examples of plants which benefit from frequent splitting and re-setting. The majority of Michaelmas daisies – reset very small pieces every year. Every three years for *Aster* × *frikartii* and the cultivars of *A. amellus*. Border yarrows are best done every year, for example, *Achillea ptarmica*, *A.* 'Taygetea', 'Moonshine', 'Coronation Gold'. Indeed, 'Moonshine' often settles down to a completely non-flowering middle age, if left undisturbed.

So too with *Coreopsis* 'Badengold'. *C. verticillata* flowers for twice as long if kept young by frequent division. Giant chives, *Allium schoenoprasum* var. *sibiricum*, too. *Rudbeckia* 'Goldsturm' is another. Young colonies give you a tremendous spectacle from early August till late October and, if kept damp, are as happy in shade, which they illuminate so well, as in sun.

On most soils, perennial lobelias need frequent rejuvenation by spring splitting, otherwise they die out in the middle, and I like a thorough upheaval on my border phloxes every fourth year. Many of them, given a warm summer, will then flower a second time in the autumn, if generously treated in this way and with irrigation.

So too with hostas. 'Never require disturbance,' we are told, but ask yourself honestly if your hosta colonies still give you the pleasure they once did. Indeed, I rid myself of 'Honeybells', so quickly did it (after a mere two years) become too congested to flower in the centre of its colony; only at the margins. If you deal with your hostas every fourth year the task of lifting and splitting is less likely to defeat you. Otherwise it'll be a question of persuading some hearty young man to perform, with fork and spade (mind he doesn't snap the shaft) what can become a Herculean engagement. Be warned, while I am on the subject, that some hostas are a great disappointment in the year following an upheaval. The foliage of *Hosta crispula*, for instance, becomes unrecognizably emaciated, the marginal white band pathetically meagre.

Lupins and delphiniums are at their finest in the year after young plants have been raised from spring cuttings or from spring-sown seed. Peonies will carry on undisturbed for many years but eventually their gnarled old crowns become largely blind and unproductive. These tough old roots are not much use for propagation. Detach each resting bud with a small piece of tuber and line out for a season before replacing an old colony with this young material.

Baptisia australis is another fleshy-rooted perennial that goes on and on so that you're almost unaware of its gradual deterioration. In this case you can raise fine young stock from seed. When the seedlings are a year old they can be lined out, and a year after this they can be moved to their flowering positions. This deep, indigo-blue-flowered relative of the lupin has great distinction.

Hemerocallis, the daylilies, suffer greatly from being dubbed as no-trouble plants that can be left undisturbed indefinitely. The modern hybrids can become very shy and benefit from replanting every fourth year. Those, like *H. lilioasphodelus* (syn. *H. flava*), having an early season, can be cut to the ground after flowering. This prevents their superabundance of lank foliage from digging their neighbours in the ribs or flopping sleazily over a path. The shorn plants quickly refurnish and may even flower again a little in the autumn.

Almost every species of iris seems to have its variegated-leaved mutant. When grown entirely for foliage effect, most of these are at their freshest and most radiant in late summer and autumn, following a spring planting.

Grasses come in apropos. They should be replanted in spring. Old colonies of clumpy ferns, notably *Dryopteris, Athyrium* and *Polystichum* cvs, are spoilt by congestion, as the whorl of fronds that composes each crown is best appreciated if seen as a unit, but becomes unnecessarily confused when a large number of such crowns have interlocked.

A principal merit of the pure white *Viola cornuta* Alba Group, of which I so often sing the praises, is its continuing season from May till November. But this continuity depends on cultural treatment. The plants must be well fed and watered, and I find they need rejuvenating by division every third year. If you can be bothered to split and replant *Campanula poscharskyana,* and its superior cv. 'Stella', after flowering (at which time you should grasp and wrench away all its flowered stems), a second crop of blossom will be put forth in the autumn.

Bergenias need attention every third year at least if they are to flower as they should. A variety like 'Morgenrot' will give you a second crop if you replant it after its spring flowering. Heucheras and heucherellas, at least on my soil and in my garden, turn woody and unproductive very quickly, unless handled every other year.

I would never recommend anyone to try and move or disturb *Alstroemeria ligtu* hybrids, even if they were changing gardens, but the fact remains that old colonies can become thin and gappy, especially if there is competition from neighbours, as is inevitable in a mixed border. My method of injecting a youthful elixir here is to raise batches of seedlings in 5-in. pots and to drop these, when a year old, into the gaps.

2

MAINLY WOODY

AFTER THE ELMS

Elms have died by the million and are still dying. Many will never be replaced but many, thank goodness, will, and especially where they come close to or within the boundaries of our gardens. What should or could these replacements be? I intend to nibble at this juicy subject, any excuse for writing about trees being welcome.

But let us not wholly dismiss the elm at the outset. I think it has, in my lifetime (for I was schooled in the Midlands where it is most abundant), given me as much pleasure as any tree I can think up. I am not so keen on the wych elm, *Ulmus glabra*, (a) because every May when its seeds ripen it looks as though it is dying, (b) because its leaf is too big and undistinguished, (c) because the tree itself, though easily recognized at a distance, is somewhat lacking in personality. The wych elm is in any case not in such dire danger, for its range as a native and freely self-sowing tree is in the north, where Dutch elm disease is neither so rife nor, thanks to the cooler climate, likely to become so.

It is the stately English elm, *U. procera*, with branches rising in billowing folds like a great cumulonimbus cloud, that has been dealt the dirtiest blow. Yet this tree has one reserve of exceptional resilience in its capacity for suckering. The tree may die, some of its suckers will also die, but never all; never, never. Sooner or later the disease will spend itself, become less aggressive and subside. The elm suckers will be ready to grow into trees again. Not that, with modern hedge cutting and grubbing practices, they will often be allowed to return, but some will. All the same there is much replacing needed.

The question of costs, when planting trees, is of primary importance. If substantial numbers of a particular species are required we should seek a supplier who offers comparatively little choice but is able, by concentrating on many of the few, to maintain a very reasonable price. James Smith of Tansley, Matlock, Derbyshire, is notable in this respect, but this firm concentrates even more on shrubs than on trees and one might turn to a nursery whose main trade is in forestry but which has also considered the needs of the ordinary retail customer. Such, in my part of the world, is Oakover Nurseries, Ashford, Kent.

If a tree is required for a key position or if we are tree enthusiasts, happy to be carried away by the excitement of planting and looking after something uncommon, then we shall be prepared and, indeed, shall have to pay a lot more. For the most patient of us the possibility of raising our own trees from seed should not be discounted. Where species are in question and fresh seed available, this can be much quicker than is generally supposed. I planted a pear seed (*Pyrus amygdaliformis*) two springs ago, and it made such a strong plant that I was able to plant it out in my orchard eighteen months later. Do beware damage by rabbits and hares, incidentally. In a night they can kill a young tree by nibbling the bark off all round its stem, and yet a timely circle of netting would so easily have prevented this. I've still not learnt this lesson myself but I like to think optimism has its compensations.

I suppose the most obvious replacement for an elm would be an oak or ash, our basic forest fare. Personally I should avoid an oak except that I love growing them from acorns. There are two native species and many hybrids between them, and one does see a good many of them around without perhaps feeling a strong urge to add to their numbers. Many, on the other hand, are past their prime, so perhaps there is a need, since most tree planters are impatient for results and our native oaks need time.

They are always recommended as the ideal providers of overhead shade in woodland gardens where rhododendrons predominate. The oak's roots are deep and do not compete much for nutrients with the shrubs beneath them. Furthermore, their fallen leaves are not large and obtrusive, and they rot down to mould in a reasonable space of time. The great snag is that the honeydew that falls from their leaves in early summer, when aphids are feeding on them, coats the rhododendron and other leaves beneath them, and this sticky-sweet substance is then colonized with black sooty moulds which make the young rhododendron leaves hideous and prevent them from functioning properly.

I have asked a few rhododendron specialists if they have any suggestions for alternatives to the common oaks which combine their virtues without including this fault (some leaves, after all, are distasteful to aphids), but so far none have been forthcoming. Another essential in this kind of overhead shade is that it should not be too heavy.

An oak that is cheaply available in quantity is the North American *Quercus rubra* (syn. *Q. borealis*). It is also easily raised from seed, and Messrs Barilli and Biagi of Bologna, Italy, offer acorns at a tiny sum per pound. This is a large-leaved oak of singularly fresh yellow-green colouring in spring, and a marvel at that season when partnered with a purple beech. It is really fast growing on decent soil; too fast for its strength sometimes. It is handsome in autumn too, but takes on nothing like the fiery tints of *Q. coccinea* 'Splendens'. *Q. coccinea* is the North American scarlet oak, but it is generally raised from seed, and seedlings vary greatly in the amount of colouring they take on in autumn. *Q.c.* 'Splendens' is a clone whose flamboyance can be relied upon, but it has to be grafted and you have to pay for it. Tree for tree I prefer *Q. rubra*.

If it is only one specimen you are seeking then be a little adventurous. After all, Hillier's of Winchester list dozens of oaks, and we should take advantage of this unique quarry, for who knows how long it will last? One I long for when I have the space is the pin oak, *Q. palustris,* with graceful semi-weeping branches and a distinctive cut to its leaves, which are not as large and coarse as in the last two species.

I have lately acquired a golden-leaved variant of *Q. rubra* called 'Aurea', on the strength of three specimens seen in the Valley Gardens of Windsor Great Park where they made an unforgettable impression (especially as I photographed them – that wretched tyrant, the camera, has undeniable merits) seen against a blue sky at the end of May. The gold-green colouring is retained more or less until August.

Never be tempted to plant the Turkey oak, *Q. cerris*. True, it grows at great speed, but it is thoroughly unsound in wind and limb and is seldom much to look at anyway. It was greatly overplanted by our fathers and grandfathers and we are now paying the price. Turkey oaks are falling all about us. Their timber is useless; I suppose it can be burnt.

Let us turn to ashes.

Difficult to write with impartiality, here, since the common ash, *Fraxinus excelsior,* has always been my tree, but for no valid reasons that I can produce like change for a £. There is the steadfastness of its cool grey colouring, carried right through the tree as far as the black buds at its branch tips. Ash twigs are thick, the tree looks strong and yet graceful too. In a well-grown, free-standing specimen (ashes need deep, well-drained yet water-retentive soil), there is a great arching sweep of its lateral branches up and away from the trunk and then down again, even to near ground level, if that is allowed.

Its leaves are ever late to flush: an ash seldom looks clothed before the third week in May, and then comes that extraordinary transformation from the severity of its winter aspect, carried forward till spring is almost gone, to the lightness of its pinnate foliage that never casts a deep shade.

A battalion of ashes was planted on our garden's west side to screen us from the prevailing wind, which they do. Their seeds also blow in on the prevailing wind. Although not as pestilential as sycamore seeds, they are a nuisance. At various times I have considered thinning the plantation by eliminating the female, seed-bearing element, for in the ash the sexes are usually on separate plants. However, some of the best units are females so I've never had the strength of mind.

The great advantage of ashes on the west side of your garden is that you can enjoy the silhouette of their foliage against the gloaming of the evening sky. It gives you a marvellous feeling of buoyancy. It's funny how you can identify yourself with a tree on occasions like this. The tree looks buoyant and thus you yourself feel buoyant. Very likely it was for reasons such as this that the society calling itself The Men of the Trees was saddled with its mirth-provoking title.

The weeping form of *F. excelsior* 'Pendula' is the most beautiful of all weeping trees. Can be, I should say, and is when it develops a height of 30 or 40 ft and an angular outline of its main branches that resembles the trees in Chinese painting. But some weeping ashes look like tea cosies and never grow out of this stage. I once asked Alan Mitchell if there were different weeping forms of the common ash but he said not. If you do acquire this wonderful tree it would certainly help (as with weeping birches and beeches) to tie a long pole to its trunk and train a central branch against this so that the weeper gains height to weep from. Grafting high on the stock would also give it a leg up, but that is the nurseryman's business. Bean tells us that the specimen in the Glasnevin Botanic Garden (outside Dublin), mentioned by John Loudon, must have been grafted at about 30 ft, judging by the picture of it made in 1838. What is probably the same tree was 50 ft high in 1966, he adds.

The manna ash, *F. ornus,* is the type of all so-called flowering ashes, and can be purchased even more cheaply than the common kind, for it comes easily from seed. All ashes flower, but in these species the blossom, which expands in May, is conspicuously white, instead of the usual purplish maroon. The scent is said to be faintly disagreeable; it is some time since I sniffed around a manna ash in flower, so I cannot add my personal interpretation. For small gardens the Ornus-type ash to acquire is *F. mariesii*, which grows little larger than a bush and flowers regularly.

The common ash is one of a number of trees which sheds its leaves in southern England without in any way colouring up first. In fact they usually fall green. But in the north country and in Scotland many ashes (not all) change to clear yellow and are radiant in autumn sunlight. It would be interesting to know the physiological reason for this difference, shared also by Norway maples and horse chestnuts.

The horse chestnut, *Aesculus hippocastanum,* is the most exciting flowering tree we have in Britain. There is no other of anything like comparable size that makes such a spectacle of blossom. And this is matched by young foliage of sparkling green colouring and intensely three-dimensional form. As a small boy I can remember being overwhelmed by the luxuriant beauty of a vast flowering horse chestnut in Dumpton Park at Broadstairs, where our school playing ground was, and this species has always made me catch my breath at the moment of its sudden, yeast-like leavening into an expression of all that is most spring-like about spring.

The red-flowered chestnuts are a little later flowering and cannot therefore be used with the white to act as a foil to their colouring. Being coloured they should be even more exciting, reason tells us, and yet they are but a shadow of the white, less ebullient, less brilliant, less everything except expensive. If you must have one, make it *A.* × *carnea* 'Briotii', which has the strongest colouring.

If you are planting a row or group of white horse chestnuts, do remember what is so often forgotten: the size they will grow to, and do allow for this because it is a shame to have them jostling each other when still less than half grown. You should allow 100 ft between plants.

There is a strong case for growing the Indian horse chestnut, *A. indica,* as an alternative. Probably it will only make a tree of half or two-thirds the size, in this country, but it grows quickly, and its (basically) white flowers are freely produced at the most welcomely late season of June and even into July. It is easily raised from fresh conkers.

The other kind of chestnuts, called sweet or Spanish, are related only in their English name, and actually belong to the beech family. *Castanea sativa,* to give the common species its rather beautiful botanical name, makes a wonderful tree, getting better and better as the centuries pass so that by the time its trunk is of an enormous girth and hollow, a specimen is reaching its climax of individuality. The fantastic feature in a very old tree is the way the parallel lines in its deeply furrowed trunk take on a spiral twist, travelling once or more round the entire trunk's circumference. However does this come to happen?

Spanish chestnuts are late in leafing. The leaves are long and plain, coarsely toothed and with a tremendous gloss on them. Nothing could look healthier. The flowering of trees in their prime is a considerable excitement, for it happens in late June and early July when one has ceased to expect this kind of efflorescence. The long creamy-white male catkins are loosely bunched in fountain-like sprays. The only trouble here is the strong and sickly scent they waft. I find these kinds of scent – privet is another – far less disagreeable than I once did, but even so it might be wise to site your chestnuts at a fair distance from your bedroom. Their fallen leaves are rather a bore, too, being more obtrusive than the average. All the same, Spanish chestnuts should be high on one's list of large trees to plant for ornament. They are never in trouble during periods of drought (unlike birch and beech).

I used to collect the fruit from our woodland chestnuts when I was young and enthusiastic, bury it in a tin, and then bring it out at Christmas for the turkey stuffing. But what a dreadful palaver its preparation was. And as I was expected to do the preparing my enthusiasm soon evaporated. I would as soon hand-weed a patch of bearded irises.

On to the lime. A sweet chestnut avenue is preferable to one of lime any day, but the latter is far commoner. The trouble is that these avenues were nearly all made of the most unsuitable *Tilia* × *europaea.* It has two major drawbacks. First, the unsightly forest of suckers produced from its bole; second, its martyrdom to attacks by aphids, their secretion of honeydew and the rain of this disgusting fluid on luckless beasts and mortals, their hides, fleeces, clothes and motor cars, beneath. On underplantings too, with sooty moulds quickly developing on fouled leaf surfaces. Leaf-fall in the common lime is early and as unprepossessing as a tree can be about it.

Yet I still feel deprived at not having been brought up in the vicinity of a lime. That scent of lime blossom in July is very wooing. Fortunately there are other species. *T. euchlora* has a glossy leaf and aphids shun it. Bean wrote of its pendulous, rather graceful growth. He was a staunch protagonist of this lime, and it is planted a good deal, but Alan Mitchell, who is not to be disputed on such a matter, tells us in his absolutely indispensable *Field Guide to the Trees of Britain and Northern Europe* that the crown of the tree at length becomes unattractive, with down-curving branches that thicken grossly when the tree is some forty years old. Finis to *Tilia euchlora*, except for pleaching purposes.

For general planting, including avenues, *T. cordata*, the small-leaved lime, is clearly the answer. It makes a handsome tree on which the blossom stands out conspicuously, and it is not notably popular with aphids. Furthermore it is cheap.

But if you only wanted one or two specimen limes, *T.* 'Petiolaris', provided you have the space for a tree of the largest dimensions, is the one to choose every time. Unlike so many of its kin, which turn dark and stodgy in late summer, this species is not only one of the latest in flower (July–August) and richly scented at that, but its leaves have very long stalks on which the blades – green above, grey beneath – continually twist and turn on the lightest breeze. Furthermore, with its high crown and pendulous branches, *T.* 'Petiolaris' is well named the weeping silver lime. Just to stand underneath one and look up through its branches gives you a feeling of elation. In autumn it normally changes to clear yellow, with some branches in this condition while others are still green.

Tilia 'Petiolaris' is not known in the wild and is nowadays considered to be no more than a hybrid of *T. tomentosa*, the European white lime, which shares many of the former's best characteristics and would be a satisfactory alternative should a good specimen of 'Petiolaris' be too expensive or difficult to find. *T. tomentosa* shows the pale undersides of its leaves in the same way, and does not even need a wind to move them into this position. It is similarly late flowering, sets fertile seed from which it is easily raised, and is hence not an expensive plant. Perhaps a little on the stiff side as a young specimen, but it soon luxuriates.

Lime blossom is popular with bees, but it has a narcotic effect on them so that they fall to the ground and crawl about in a drunken stupor. If you have long-nosed dogs, like dachshunds, that are for ever prodding and probing and trying to push the bees on, it is as well not to site a lime by a courtyard, for instance, where your pets are much afoot. A dachs that has been stung on the nose looks like a new comic breed, but it won't like being laughed at.

The smooth, pale greyness of trunk and principal branches gives a cathedral-like quality to a mature plantation of beeches, and this is how they can be most enjoyed – from within, especially if there is a blackbird singing, for its voice will be given resonance by these naked pillars. The ground underneath may be almost bare except for mosses, drifts of warm brown leaves, and perhaps the large white helleborine, which always seems on the point of opening wide and dramatically in late spring, yet never quite does.

One of the most beautiful sights I know is the thick, grey-green, undulating carpet of white fork moss, *Leucobryum glaucum,* underneath beeches at the entrance of the Savill Gardens at Windsor. This reaches its peak in May, when it has to vie for the visitors' attention with rhododendrons and azaleas in a Joseph's coat riot of eye-catching colours, and yet the public do notice and appreciate this subdued spectacle. I spent time one morning watching their reactions. It is difficult to start a moss garden of this kind if it has not arrived of its own accord – you would need to introduce very small pieces – but mosses will grow under beeches where nothing else will or, if it does, a competitor is easily removed.

Beeches are notorious for the depth of shade cast by their foliage in summer on the ground beneath them. They are, in fact, the most efficient of ground-cover plants, and yet too few gardeners are content to leave it at that and seem to think the ground should be further cluttered at a lower level.

Crocus tommasinianus, in shades of mauve, and the yellow winter aconite, *Eranthis hyemalis,* will thrive in company beneath a beech, as I know from the example in a friend's chalk garden at Chilham, Kent. Their secret of success is that they complete their growth cycle before the beech's leaves expand and while the ground is still damp. But it is very unwise, as the same friends have pointed out to me, to plant a beech as a lawn feature. The hard cups that enclose the beech nuts are exceedingly durable. They get pressed into the surface of the lawn and make a nightmare of mowing.

Young beech foliage, whether vivid green or copper, is among the most spell-binding of spring's contributions, and yet this tree can easily be faulted, from a purely visual and aesthetic angle. Its branch system always seems overcrowded and there is no character in individual branches. They're a bit of a mess, in fact. And a beech in high summer is among the dullest of trees.

Purple beeches – just the odd one here and there – do add variety and distinction to the scene. However sombre in themselves once their spring has settled into middle age, they remain a foil. In any quantity they would be asphyxiating, of course. They are not so very much more expensive than the type plant. Both are raised from seed, and the purples (copper, if you like) come more or less true to colour from seed for most of the time. But there is a good deal of minor variation to be expected. In habit, too. Seed provenance dictates the shape of your tree to a large extent. On the whole you would be wise not to grow on just any beech seedling you happen upon, tempting though this is, for the newly germinated seedling is fascinating, with its two large cotyledons: almost as exciting as a germinating avocado pear.

As a variant to the common beech, *Fagus sylvatica*, the cut- or fern-leaved *F.s.* var. *heterophylla* 'Aspleniifolia' makes a delightful specimen with an airiness retained throughout the summer. Alan Mitchell makes a fascinating comment on it in his *Field Guide to Trees*: 'Where branches are cut or damaged, ordinary entire leaves will appear . . . because the plant is a *chimaera* with inner tissues of ordinary beech overlaid by tissues of the cut-leaf form.'

Among the largest weeping trees at our command, a weeping beech should come high on a list of preference. It has a splendid individuality that proclaims itself at a great distance. You might plant it as a focal point to close a vista. The fastigiate Dawyck beech is in complete contrast. Like other fastigiate trees it has been much planted where space is restricted, although it bulges considerably in old age and then becomes all the more interesting. I find it difficult to recommend fastigiate broad-leaved trees that I really like, but this is one of them. A purple-leaved sport of *F. sylvatica* 'Dawyck' was exhibited to the R.H.S. in June 1973 and won an A.M. It would be hard to place this effectively in a country garden but it should look well against white harling, for instance.

I should like to see more plane trees in the country. They are so magical in London parks and squares, winter, spring and summer, that I can scarcely stop gawking at them to shift my stare on to the passers-by. Bean, in the lately revised edition, says of *Platanus* × *hispanica* (syn. *P.* × *acerifolia*): 'The London plane has a characteristic crown: the branches are somewhat tortuous, and the perimeter is intricately branched, giving a winter silhouette which is surprisingly delicate for such a robust tree . . .' That hits the mark. The other beautiful aspect of the London plane in winter is its persistent fruit: globes of varying sizes that hang in loose chains of from one to five or six links. In spring these break up into down-light seeds with tufts of hairs attached that float them on the wind until they fall with the stealth of snowflakes.

The young green of plane leaves is a marvel, retaining its freshness for a surprisingly long time. Much foliage, on some trees more than on others, is held until late in autumn, even into early December. This is when you hear the grumbles, for the leaves are large and do not readily disintegrate. It would be madness to site a plane near flower borders, but in a landscape setting or on the garden's perimeter it is another matter. I should add that there are a great many hybrid planes in London and elsewhere, and most, especially one called 'Pyramidalis', are inferior to *P.* × *hispanica* itself. Thus Bean, again. How one is to know what one is buying therefore remains a problem, as no distinctions are generally made by nurserymen.

The oriental plane, *P. orientalis,* is one of the London plane's parents. We meet it from Turkey through Iran to India as the chenar, an enormous and impressive tree with a huge upright bole and spreading branches, often forming a village centrepiece. Why, then, does it behave so differently in England, where its several trunks habitually lean, loll and layer themselves? These composite oriental plane units – such as you see at Blickling in Norfolk, for instance – are very, very beautiful in their way, the leaves being much more deeply cut than in *P.* × *hispanica,* but they are not practical trees for general planting.

Platanus × *acerifolia,* the former name of *P.* × *hispanica,* means 'the plane with maple leaves'. On the mutual back-scratching principle we find *Acer pseudoplatanus,* the 'mock plane maple', actually the sycamore; and *Acer platanoides,* the 'plane-like maple', known to us as the Norway maple. Both trees look finer in the north than down my way. I think of sycamores particularly around farm buildings in west Yorkshire. On Orkney it is the only tree generally planted for shelter, and although it looks scruffy it is absolutely hardy and does an invaluable job in the first line of defence. Except for this purpose I cannot recommend the sycamore to any gardener. Its habit of self-sowing is pestilential. But I am fond of the variegated sycamores. There is a lightness about them and this remains right through the growing season. Their leaves are streaked in green and cream in a disorganized way that does not bear close examination but they never revert. Indeed, they are almost invariably raised from their own seed.

The Norway maple, *A. platanoides,* again makes a grander feature in Scotland, where it also colours brilliantly. Its leaves are more sharply cut than the sycamore's but before they unfold in spring there is a brief yet delightful display of lime-green blossom.

The cultivar called 'Goldsworth Purple' is exceedingly heavy in its summer dress but does make an effective foil to a golden-leaved tree like *Robinia pseudoacacia* 'Frisia' or *Quercus rubra* 'Aurea'. The best-known variegated Norway maple is 'Drummondii', which has a broad cream margin to each leaf, and this is a far more distinctive arrangement than any blotchy variegation can hope to be. Unfortunately 'Drummondii' is strongly given to reverting. I do not therefore see the point of growing it as a tree, unless you fancy yourself as a monkey with a saw.

For this reason I ordered a bush specimen for myself, intending to plant it in my Long Border and cut it hard back annually or biennially. But I was sent (and charged for) a full standard just the same. It is becoming impossible to buy bush forms of trees nowadays, and the reason is that standards are so much easier for the nurseryman to manage in the field. He can get up and down between the rows with machinery in a way that bushes do not allow. We must therefore, as so often, be dictated to by the economics of the situation.

I have no wish to make this a tree book and shall wind up my subject summarily at this point. Readers who have only small gardens to manage (and they are well aware of being the majority) may already have a restive, 'what's this all to do with me?' feeling. Yet trees concern all of us, whether we plant them ourselves or not. An informed, interested public with an awareness of their importance can make its pressures felt in many ways. It is no good just grumbling when we see poor tree planting programmes, or no programme at all, going on around us. We need to put in an oar. Otherwise we shall be dominated by that large and miserable section of the community that detests trees because they are larger than we are and therefore destined, inexorably, to fall on and crush us. Come to that, I can think of some whom I could spare.

THE FLOWERING OF NON-FLOWERING TREES

Gardeners all know what they mean by flowering trees and shrubs: those that flower significantly and make a display, a show; earn their keep, pull their weight. Although they don't refer to the rest as non-flowering trees and shrubs, the assumption is that they exist. In fact, all trees and shrubs do flower (unless they are mutants that have lost the capacity), but may be too demure in their manner of setting about it to gain general recognition from a public that expects flowers to flaunt. Still, there's a deal of quiet pleasure to be derived from the flowering of non-flowering trees.

The first of them to tell us that spring is on its way is the hazel. Its catkins are male flowers and they may expand even before the new year. Perhaps this is really more of a shrub or scrub than a tree. The alder, whose catkins follow, is undeniably a tree, and among the handsomest of our natives, especially in the north country. It may flower before the end of January but individuals vary widely in their timing, not just according to season but one as against another, with some coming into full flower in mid-March. They glow with a rufous warmth as the catkins expand, changing to muted yellow as the pollen is released. The grey alder, *Alnus incana,* has rather slenderer catkins, but being regularly ten days earlier with them than the earliest common alder is especially cheering. Rarely does it start flowering after the end of January. The one that really works me to a frenzy of enthusiasm is the Italian alder, *A. cordata,* although this seldom flowers before early March. Its fat, 6-in.-long catkins are splendid, and a brighter yellow than one expects of an alder. Furthermore its female cones are very substantial and survive from the previous year with devastating glamour.

One of the most dramatic releasers of pollen at that season is the common yew. A male tree, blown about by March winds, looks as though it is smoking: clouds of pollen are puffed out. My favourite early spring non-flowering tree, however, is the common elm, which is covered on every twig with a glow of genuine red that looks marvellous in the sunshine. True, this red is soft and undramatic – not to be compared with the flamboyant *Caesalpinia* (syn. *Poinciana*) *pulcherrima* – but especially subtle because of the way its blossoming emphasizes the cloud-like billowing that is so characteristic of the elm's outline. It must be unusual for an elm to win the R.H.S. Award of Merit as a flowering tree, yet this happened on 19 February 1974 when the Director of the Royal Botanic Gardens, Kew, showed *Ulmus villosa.* It has wand-like shoots that are lined from end to end with off-*white* bobbles of blossom at every node. I am told that the branches of this tree can sweep the ground and that its leaves are poplar-like. There are so many historic trees to see and admire at Kew, but when you are there, all recollection of which you should be looking at ebbs away and you are simply transfixed by the Palm House!

Many poplars, including aspens, flower in March; they are catkin bearers but, being large trees, you need to arrange your position and their backgrounds rather carefully if they are to be seen at their best. The general tone of these catkins is pussy-cat grey, but the males have rich red anthers among the grey, and can be unexpectedly exciting. The grey poplar, *Populus* × *canescens,* whose pale foliage and trunks make it a pleasant companion all through the year, has the additional merit of nearly always being male, so that you can be sure of having the kind of catkins you want on it.

With the larch, it is the embryo female cones that are alight in spring; these are crimson red and in striking contrast to the young leaves. Branches cut for the house come out well in water. As larches are monoecious, you can be sure of finding female flowers on every tree.

Flower arrangers are sharper-eyed and better aware of the potential in the flowering of non-flowering trees than is the commonality of mortals. They make frequent use of hornbeam blossom, for instance, whose green spikelets expand at the same time as their young foliage, just when daffodils and blackthorn are at their best. Hornbeams don't flower freely (if at all) in every season – doubtless because the heavy, and again decorative, seeding that follows exhausts them. Recently in south-east England they have been flowering in even years and taking a rest in the odd ones (unlike the common ash which has been doing it the other way round). Hornbeams in fruit are no less arresting, but are beaten in this department by the nearly related hop hornbeam, *Ostrya carpinifolia.* This won an A.M., unanimously, for its fruiting when shown from Kew on 13 July 1976. The popular name gives the idea: a hornbeam densely hung with hop-like cones.

I always marvel when a success is made of young oak foliage and catkins in a flower arrangement. One appreciates the work that must have gone on behind the scenes to enable this tender greenery to stand without wilting, once the arrangement has been completed. It is naturally the English oak, *Quercus robur,* that is used most often, but if you are planting an oak, it will be well worth giving consideration to the Pyrenean, *Q. pyrenaica,* whose catkins are exceptionally long and hang conspicuously on the tree which wears them.

Do consider growing *Elaeagnus angustifolia* as a lawn specimen instead of the customary weeping pear (*Pyrus salicifolia* 'Pendula'). Both have silvery, lanceolate leaves that put you in mind of a willow, but the elaeagnus (or oleaster) has masses of tiny yet deliriously scented blossom in early June. It is the scent that gives their inconspicuous presence away.

Eucalyptus an surprise you by being suddenly covered with white blossom when you least expected it. The last time this happened to me was in Sheffield Park Garden, Sussex, which I was visiting on a perfect day in late October for their famous autumn colours. Their enormous *E. gunnii* were in full bloom. In fruit also; I was allowed to take some capsules and grew a nice batch of seedlings the next year.

One reads about and sees pictures of startling red blossom on some gum trees in their native Australasia, but it unfortunately appears to be the case that these colourful species are among the least hardy and not for us in Britain.

EVERGREEN BROADLEAVES: THE FEW PRECIOUS TREES

I would as soon see a deciduous tree in winter as an evergreen, but the fact is that, deciduous trees being the rule in Britain, many gardeners hanker after evergreens. Their natural choice where a tree is required will be a conifer, because so many conifers are hardy. But there are a few broad-leaved evergreens of tree-status worth considering.

Most do not grow very large, and most are on the borderlines of hardiness and will require a reasonably sheltered position. They have a tendency to shed their foliage at seasons (especially late spring and early summer) when we feel that we should not have to be coping with fallen leaves, so you must make up your mind in advance of planting not to be cross with them for this habit. Neither must you hold it against them that their leaves, being tough and leathery, are slow to disintegrate in death.

It is true that the bay laurel, *Laurus nobilis,* has the soul of a bush, being many-stemmed and a natural suckerer, but it attains the dimensions of a tree. There was one here before the garden was made. It is close to the house, I see it every day (even as I write), and I love it. Its yellow-green colouring is marvellously cheerful in winter. Bad winters hit it back severely, but it always recovers. When its branches are bowed with snow to the ground it reminds me of Shepard's pictures of Eeyore in *The House at Pooh Corner.*

The strawberry tree, *Arbutus unedo,* is quite as hardy. It is easily raised from seed and is apt to be an up-jumped bush, though its rufous, scaly trunk and main branches become impressive in time. It has the happy knack of flowering and ripening its previous year's fruits simultaneously in the autumn. Red and white together. Red admiral butterflies sup at its nectar. Not showy but pleasing. Whether your specimen develops the fruiting habit or not seems to be a matter of luck. *A. menziesii* flowers in the spring; it rarely fruits in this country. The great thing about this species is its smooth trunk flushed pink and yellow. It hates being transplanted, and I have failed with two seedlings I raised.

Drimys winteri is not all that hardy but probably hardier than most of us realize. We haven't got around to trying it, and it must be acknowledged that you don't exactly trip over them at garden centres and other places where you hope to buy shrubs. It makes a tall slender shrub-tree clothed to the ground with luxuriant lanceolate but broad-ended foliage. Scented white flowers are borne terminally on the shoots in July and they make quite a show. There is an enormous specimen at Inverewe, in N.W. Scotland, which makes your mouth water when flowering.

I am not sure whether it is fair to write of evergreen magnolias as trees in this country, but I have a friend in whose London garden *Magnolia delavayi* has developed into such a beautiful specimen, now 30 ft high and with a tree-like stem, that I think other gardeners, similarly favoured, should give it a trial. It has enormous leaves of sea-green colouring and although the flowers, borne intermittently through the summer from quite an early age, don't last long, they make a contribution – they have the distinction as well as the colouring of chamois leather, but additionally a powerful scent.

Many rhododendrons make trees in time, and when the climate suits them, but the one species that best qualifies (as witness its name) and that I want to push is *Rhododendron arboreum*. Narrow for its height, it does make a marvellously shapely tree offset by a splendidly rugged trunk. The flowers may be white, pink or red and all are good. The only thing is, it flowers desperately early, never later than April, so it needs shelter more than most if frost is not to deprive you of your pleasure.

Quercus ilex, the hom oak, but known as ilex *tout court* by its intimates, can grow vast in the south. There is an immensely impressive specimen in the next village to us, behind which a house cowers in permanent gloom, but its American owners fortunately appreciate that the tree is more important than the house and them put together. The ilex is a beautiful, melancholy olive green. It stands up to salt gales but is not for a really cold inland climate. I wish I saw the cork oak, *Q. suber,* more often. It is always a thrill with its rough, so obviously corky trunk, and it will thrive in most sheltered gardens. Again, its leaves are olive green.

We should not despise our own holly, *Ilex aquifolium,* just because it is wild and common – less common since mechanical hedge-trimming came in. Even the males make fine trees with smooth, whitened trunks, and their white flowers are abundant in May. If you want just one do-it-yourself holly that berries without worrying you with its sex problems, then *I. aquifolium* 'J. C. van Tol', alias 'Polycarpa', is for you. No prickle problems here, either, nor with *I.* × *altaclerensis* 'Camelliifolia' – a female with enormous, long lustrous leaves on pendulous shoots, and an abundance of berries if a husband is lurking in the vicinity.

If this has camellia-like leaves you are just as likely to be deceived in the same way by a splendid evergreen privet, *Ligustrum lucidum*. It runs up to 30 ft with ease, and its great panicles of white flowers are most welcome in their September season. Do grow this one even if you have a thing about privet.

A despised shrub that is spoilt if, as is usual, it has to be kept as a shrub or hedge by clipping (under which ignoble treatment it readily succumbs to the silver leaf fungus) is the extraordinarily tough and hardy Portugal laurel, *Prunus lusitanica*. As an untrammelled tree it is super, and its cascades of white blossom on drooping spikes in late spring are a thrill.

Gum trees, of course. These can grow very large but should be planted very small (say a foot high) in spring and before they get pot-bound, otherwise you will have protracted troubles apropos of wind-firmness. I am glad that gum trees are now being much more widely planted than could ever have been expected even fifteen years ago. Not only is their grey or glaucous foliage attractive both in youth and maturity, but their smooth, peeling trunks are a marvel. There are a good number of large-growing hardy species – hardy, at least, until the next severe winter catches some of them out – but still a lot remain hardy in many places. With its pale stems and branches (though variable in this respect), *Eucalyptus pauciflora* subsp. *niphophila*, the snow gum, is one of the prettiest smaller-growing species. It always has a lean on it, Roy Lancaster told me when on a visit. A few days later I met a bolt-upright specimen in Patrick Synge's West Sussex garden. It was interesting to happen on the exception that proves the rule so promptly. The leaves of this species have an extraordinary metallic glint to their surface which you only notice when sunshine catches them and sets the whole tree sparkling. You think they must be wet but this is not so.

Finally one of the southern beeches, *Nothofagus*. These are related to the beeches but, unlike them, are calcifuge. Otherwise they are extremely useful and quick-growing trees, the commonest of them deciduous, but *N. dombeyi* is a good evergreen, its small oval leaves having toothed margins. Anywhere sheltered in the south of England this tree can be expected to attain 50 ft in as many years, which is fast enough for the size of most gardens today.

AMONG THE CRABS

Crab apples are so called because of the sour, harsh, tart, astringent quality of their fruits (thus the dictionary). We also speak of ornamental crabs in the way that we do of ornamental cherries. The latter are grown entirely for their flowers, but the former can be admired either for their flowers or for their fruit or for both, in their seasons. And our sense of enjoyment and admiration will reach its highest pitch if we know that, to crown all, we shall be able to make delicious jellies from these beautiful fruits. For this purpose we *need* them to be harsh and astringent, for therein lies the interest and excitement of the final product, in notable contrast to the boring blandness of jelly made from domestic culinary apples.

The crabs are a large and complex group, and it is unlikely that we shall have room for more than a few of them in our gardens. So we need to ask ourselves, before making a choice, what our principal requirements are. Shall we be content with blossom that lasts only a week or ten days and has nothing to offer thereafter, not even a tree of pleasing shape? Put like that the question would obviously receive a negative reply, and yet that is the precise choice that many gardeners make, carried away as they are by a glossy catalogue picture or by an exhibit at a flower show or by a tree seen in its moment of glory.

By contrast, some crabs retain a spectacular display of fruits on the tree right into the new year and beyond. Such is *Malus* × *robusta*. Its clusters look more like rosy red cherries (after they've been dyed for use in tinned fruit salads) than apples. I remember my mother once taking a fancy, as well she might, to some in a table arrangement at a luncheon party and she was allowed to bear them home. To the last she had a passion for growing trees from seed, and these had duly to be sown and grown, which they did with astonishing speed. I planted one in the orchard and it soared to 15 ft in no time, but the fruits, when they came, were quite large and green and exceedingly scabby. Not only is *M.* × *robusta* itself a hybrid, but it could have crossed with any domestic apple to produce the pips we sowed. Much cannot be expected of these frolics.

However, there are some excellent cultivars derived from and closely resembling *M.* × *robusta* in its welcome habit of holding its fruits on bare branches. *M.* × *schiedeckeri* 'Red Jade' is one such – and what a pleasing name. The really weeping crabs are a bit of a dead loss but this one makes an umbrella of a charming habit. It has pink buds and white blossom. *M.* × *robusta* 'Red Sentinel' is deep in fruit colouring and persistent in the same way. The lineage of 'Crittenden' is uncertain, as the original seedling was imported from Japan in 1921, but it won its Award of Merit when shown by Mr Ben Tompsett, who lives at Crittenden (he is a high-powered Kentish fruit farmer), on 24 January 1961. The midwinter date suggests that it has at least one chip off the *M.* × *robusta* block. Its fruits are up to an inch across and deep red – at least on their sunny side.

Now all these crabs and especially *M.* × *robusta* itself are frequently but misleadingly known as Siberian. But the true Siberian crab is only one parent of *M.* × *robusta*. This is *M. baccata*. *M.b.* var. *mandshurica* has both an Award of Merit and a First Class Certificate, and is the one to get if you have room for a splendid, wide-spreading tree, bigger than any other apple I have so far met. It is absolutely smothered in white blossom in May and a terrific sight then, but its fruits are minute and of no account.

From Russia around 1891 came *M. niedzwetzkyana* (forget it), which became the prototype and parent of a huge range of hybrids heavily impregnated with anthocyanin pigment. Leaves, flowers, fruit, and even the young wood itself are thickly suffused with reddish purple. Of the twenty-one crabs currently listed by Notcutt's, no fewer than nine are of this type.

At its best, when a tree is just coming into young leaf and flower, these are good: strong and bold. But they very soon lapse. The flower colouring fades and looks tired, raddled; the leaves take on an indeterminate greenish purple tinge, and scab moves in to make them and the fruits thoroughly sooty. The fruits themselves are often sizeable, but when of a heavy purple colouring they might just as well not be there for all you notice. I am not at all against purple-leaved trees and shrubs, as are some of my colleagues, but I do think the purple crabs are among their least distinguished representatives.

In case you disagree let's take a look at a few of them. The three *M.* × *robusta* cvs. that I was brought up on were 'Eleyi', 'Aldenhamensis' and 'Lemoinei', and I could never be sure of which was which, though the last seemed to me the best in flower. Its fruits are far too dark a purple, however. 'Neville Copeman' is a seedling of 'Eleyi' and an improvement: very free with showy globular red crabs in September. I don't care for its habit but it is good of its kind. *M × moerlandsii* 'Profusion' is an improvement on 'Lemoinei', perhaps, though its flowers quickly fade. *M.* × *m.* 'Liset' is said to hold its colour better. 'Simcoe' has Awards of Merit both for its flowers and for its fruit. The leaves are purplish but the flowers are pink and the fruits purplish red but not so dark as to be ineffective. Why saddle it with the unattractive name? Crabs vary a lot in this respect. My gorge rises to the thought of ever buying *M.* × *atrosanguinea* 'Gorgeous', from New Zealand. Just imagine if my greeting 'Hullo Gorgeous' (one wants to keep one's plants in a good humour) were to be overheard by some garden visitor.

To finish off the purple fraternity, there's a weeping one called *M.* × *gloriosa* 'Echtermeyer', described in Hillier's *Manual* as graceful, low, wide-spreading with weeping branches and purplish or bronze-green leaves. Flowers rose-crimson, deeper in bud, followed by reddish-purple fruits. But a note on it in Bean says, 'Mr Hillier informs us that he has found it to be a plant of poor constitution and very subject to scab.' It's wretched when you have to try and sell plants you don't believe in.

Some crabs have yellow fruits, and among these the splendidly named *M.* × *zumi* 'Golden Hornet' is probably the finest, with abundant crops of inch-long, oval fruits hanging along the branches' undersides. This has an F.C.C. Of similar shape but half as large again is the shining scarlet 'John Downie', which is arguably the best of all crabs for jelly. Its flowers are pink in bud, expanding to pure white, and the fruit ripens in the second week of September, soon falling off. It is subject to scab, like the majority of apples of all kinds. Of course you can always spray . . . Ten or twelve times in a season is what the commercial growers find adequate for a clean crop.

A fruiting crab that I should mention, because it was planted by Vita Sackville-West in front of Sissinghurst Castle and still makes a prominent feature there, is 'Dartmouth'. Hung with large purplish red crabs, it is an eye-catcher in a good year, but in hungry years its flower buds are regularly stripped by bullfinches.

Malus floribunda was planted in this garden before I was born and I still love to see it in late April, its crown rising above an 8-ft wall. It makes rather wand-like branches which are excellent to cut and arrange with white magnolia blossom, and they are deep pink in bud, fairly quickly fading to a pale pink when open. The tree fairly seethes, and if the flowers are on the small side this has the result of avoiding the slightest hint of coarseness and they could certainly not be more abundant. The fruits are nothing; it's just for blossom. *M.* × *atrosanguinea* used to be given as a variety of the last and looks like a deeper-coloured version, though less effective.

I've reserved my favourite till almost the last: *M. hupehensis*. Those having the irresistible urge to grow apples from pips are safe with this one. Although it is a sterile hybrid (I paraphrase from Bean) it is apomictic: that is it can produce seed asexually – a case of immaculate conception in the vegetable kingdom. All the seedlings come true *M. hupehensis*, no matter how many lechers may be hanging around. It makes a most beautiful tree, tall for its width and developing a fine trunk with large scales. This is one of the last to flower, at the turn of May and June, and its blossom is prodigious – usually pure white, but pink-flowered variants occur and are available. The crabs are small but large enough to make a show. I do like fruits that give variety by colouring on the sunny side while remaining green in shade. This is one such.

I must bring in *M. trilobata* since I have a young plant of it. Botanists would like to hive it off into another genus, and to the most untutored eye it looks extraordinary, having regularly trilobed leaves more like a maple's than an apple's; its autumn colouring is like a maple's too – bright red, even on my clay soil, normally reluctant to induce colour in any shrub or tree.

Perhaps I should squeeze in a nasty word against *M. sargentii*. This makes a bush, not a tree; a huge bush but still a bush, and quite without character. For just one week in the year, around Chelsea Flower Show time, it redeems itself with pink-flushed buds in profusion, opening white. Its fruit is insignificant.

SHRUBS IN MIXED BORDERS

That we should not exclude shrubs from borders usually styled 'herbaceous' now seems obvious to most gardeners. After all, what's the point of applying any such restriction? Shrubs have a nicely stiffening effect among groups of plants; their foliage, sometimes evergreen, contributes additional textures, patterns and colouring. The herbaceous plants, on the other hand, help by their proximity to mask any ungainliness in form to which the shrubs may be subject. The gain is in both directions.

However, there is one drawback to the mixed border: it is more difficult to strike the right balance with its contents than is the case with an all-shrub or all-herbaceous border, or a border consisting mainly of bedding plants and annuals. The management factors in a mixed border are necessarily more complex; there are greater chances of one type of plant being swamped by another and greater vigilance therefore needs to be exercised. Herbaceous material has a Jack-and-the-beanstalk way of rising in a matter of weeks rather than months from nothing to voluminous luxuriance. Any shrub neighbour must be quick enough off the mark itself to be able to cope with this. If it is small at the time of planting and slow-growing thereafter, there is a real risk that it may never make the grade. You will be under the continual necessity of keeping it free from weeds and overpowering neighbours. There it will sit, cowering in its hollow beneath lush, encircling giants. However solicitous for its welfare you may be, the chances are that you will sooner or later take a few weeks off in the growing season, and the poor wee mite will go under.

Osmanthus delavayi might be such a victim, or one of the forms of Japanese maple, *Acer palmatum*; *Elaeagnus pungens* 'Maculata' is desperately slow in its early years, and so is the not dissimilar holly *Ilex* × *altaclerensis* 'Golden King'.

Certain shrubs are far better fitted to associate with plants than others. I think it is fair to assume (anyway I'm going to assume) that the border will be required to look its best from June to September. Most shrubs flower before this. 'All right,' you may say, 'I'm extending my border's season.' But if a spring-flowering shrub is to be a passenger for the whole summer, that's not really good enough. Most rhododendrons and azaleas would therefore be excluded from the mixed border with a summer season, though one might make a few exceptions: for instance, of the azaleas *Rhododendron arborescens* and *R. viscosum*, with a particularly late flowering season; also of some whose interesting foliage would amply compensate for flowers.

Or take the mahonias. The most popular in this tribe is undoubtedly the winter-flowering, lily-of-the-valley-scented *Mahonia japonica*. In summertime, however, we have to depend, for this shrub's contribution, entirely on its foliage. Some people consider this to be a handsome feature and it does have the interest of the pinnate form with spiny margins, but it is a singularly dull green with a lifeless surface. *M.* × *wagneri* 'Undulata' flowers in early April with many other shrubs and has no outstanding quality in its pleasant scent, but the shrub is really handsome at every season, with highly polished leaves, agreeably crimped and catching the light at different points on their surface. They turn bronze-purple in the winter and the young foliage is pale copper changing to tender green.

This shrub is hardy anywhere, though it needs protection from cold winds in winter if its foliage is to remain unscorched. You must have shelter (and it does well in London) in order to revel in *M. lomariifolia*. Its upright, stemmy habit enables it to rise above and queen it over other plants, with rosettes of exquisite foliage, each leaf consisting of up to forty leaflets, curled at the margins and a light fresh green. The flowers, borne in upright clustered spikes, come in November but have no scent. You can't expect everything.

We shall not, of course, be so fussy about the foliage of shrubs that flower during our border's main season. Philadelphus bloom in June and July, so never mind their very ordinary foliage. However, where we can get some life into this feature, let's grab it, as in the yellow-leaved variety of the common *Philadelphus coronarius,* called 'Aureus'. The flowers are there too, and so is the scent, so we've nothing to lose.

Most of the weigelas flower in May and early June, and their foliage is, in the main, spitefully boring. Even the purple-leaved *Weigela florida* 'Foliis Purpureis' looks as though it had been bathed in soot water. 'Eva Rathke' has the redeeming feature of carrying a succession of its deep red funnels over a long season, but the green and yellow foliage of *W.* 'Florida Variegata' lifts this cultivar on to an altogether more exalted plane. For the average-sized garden it is the only weigela worth considering, and it blends admirably with orange and bronze flowers in a mixed border.

Perhaps the most suitable of all shrubs in a mixed border are those that can or should be pruned quite hard at regular or irregular intervals. Over these the gardener has the maximum of control. He can check or give rein to their exuberance according to the demands of the situation.

There are, for instance, the buddleias. The more I see of them the more important do I realize it to be that they should possess more qualities than one. Habit, leaf and flower truss all need to be pulling in the right direction. 'Glasnevin Hybrid' is one such. The leaves are elegantly narrow (not broadly coarse), the stems and flower spikes are exceptionally slender, and the shrub has a spreading habit; not upright and leggy as is too often the way. 'Lochinch' is constructed on more traditional lines but is excellent in its way, with clear lavender spikes offset by grey foliage. And *Buddleja fallowiana* 'Alba' has the palest foliage of all, and it blooms with commendable persistence from July to October.

Other obvious mixed border candidates are *Hydrangea paniculata* cultivars, but the *H. macrophylla* hybrids also; *Hypericum* 'Hidcote' for its flowers and *H.* × *inodorum* 'Elstead' for its red berries; *Indigofera heterantha* (*I. gerardiana*) – so much hardier than is generally realized; the shrubby potentillas; *Sambucus nigra* subsp. *canadensis* 'Maxima' (for its white flowers) and *S. racemosa* 'Plumosa Aurea' (for its leaves); the tree mallow, *Lavatera olbia*; and sundry fuchsias.

Shrub roses are ideal for mixed borders. Their rounded blossoms associate wonderfully with irises and delphiniums, acanthus and regal lilies; with lavender-flowered hostas such as *Hosta ventricosa* or *H. rectifolia* and with the white bells of *Galtonia candicans*. The amorphous or ungainly habit of the rose plant will here be absorbed and pass unnoticed. Provision should, however, be made for the fact that the looser growing roses get bowed down by the weight of their blossom unless adequately supported, and this support is most easily given in the dormant season.

Many grey-leaved shrubs are good mixed border inmates. *Phlomis fruticosa,* the Jerusalem sage, is one of my favourites. It does get large and sprawling with the years – my specimen is thirty years old with a 12-ft spread, but only 3½ ft high, because the snow breaks off the taller, older branches from time to time. *Helichrysum splendidum* is very reliable and can be tidied up with a hard pruning every two or three years.

The grey-leaved willows could be put to wider use, since their hardiness is above suspicion and that is more than can be said for the majority of greys. *Salix alba* var. *sericea* is by no means embarrassingly vigorous, and should be pruned back to a stump each winter so as to restrict its maximum growth. Its lance-shaped leaves gleam like silver fish. *S. exigua* has even slenderer, whiter foliage on an elegant shrub that will grow 10 ft tall if you wish. It has the unusual attribute, in a willow, of suckering. *S. lanata* is too slow-growing and precious to want pruning at all. Indeed, in a mixed border, it is one of the shrubs that will need protecting in youth from its associates. Of the same 3- to 4-ft bushy habit, *S. helvetica* is more typically lance-leaved, for in *S. lanata* the foliage is rounded.

Some shrubs provoke the favourite riddle question, 'When is a shrub not a shrub?' The R.H.S. Dictionary describes *Salvia interrupta* as a perennial herb, woody at the base. As this woody base is up to 3 ft tall, it looks distinctly shrubby to me. The leaves are grey-green, felted and typically sage-like. On flowering, the plant dramatically doubles its height with a widely-branching inflorescence, and its substantial flowers are 1½ in. long, of a rich purple colouring with white flecks.

The dictionary nicely describes this plant as hardy or half-hardy. It can take its choice. I am happy to say that it frequently turns out to be hardy where I thought it would be tender, although it may be killed by a soil-borne fungus disease. There is nothing else like it among border plants. It has the stemminess of *Verbena bonariensis* but above a substrate of evergreen foliage.

Ceanothus, Caryopteris and *Ceratostigma* all combine the rare attributes of being blue-flowered shrubs suitable for mixed border use. *Ceratostigma willmottianum* is another shrub/herb, its behaviour depending on climatic conditions. Following a frosty winter, the plant dies to the ground, but will retain all or a proportion of its old wood if it was mild. Never cut your plants down automatically in winter or early spring. Always wait till May to see which way the cat has jumped, and then remove only those old stems that have died. The rest will start blooming in July and continue into late autumn. But the young stems thrown up from ground level will often not get into their flowering stride till well into September. The flowers are carried in clusters and are deep blue, like a plumbago, associating excellently with red fuchsias. I cannot think why I have never put the two together. In fact, mine is behind a group of the muted red *Sedum* 'Herbstfreude' and that looks good too. Actually, blue goes well with almost any colour except other blues. Then it sulks.

The deciduous ceanothus are scruffy shrubs, but there is no need to be aware of this, if you have them emerging from a luxuriance of surrounding vegetation. My 'Gloire de Versailles' (the common powder-blue sort) was behind a bronze-flowered hemerocallis, and they made good company until I tired of the daylily (a modern hybrid called 'Helios') for being shy-flowering and when that was gone, the ceanothus died in the floods that succeeded the 1976 drought.

The hybrid *Caryopteris* × *clandonensis* has several clones of specially intense colouring. There is little to choose between 'Kew Blue', 'Heavenly Blue' and 'Ferndown'.

You could achieve a pleasing blue and yellow combination with any of the afore-mentioned trio and the shrubby *Hypericum* 'Rowallane'. It is not as hardy as the more serviceable 'Hidcote', but a great deal more striking, with deeply cupped, waxy blooms of the intensest yellow. The plant needs shelter and is apt to die back a good bit in the winter. Again, delay tidying it up till late spring. Except in very mild districts, the shrub has no distinction of form and is best grouped, several together, to make one substantial unit, and you will not generally be far out if you expect it to reach a height of 5 ft.

ROSES TAKE THEIR PLACE

What place, I often ask myself, should roses take in our gardens? A silly question, really, because each of us has to make up his mind about it for himself and each will reach a different conclusion, right for him, wrong for his neighbour.

Let me be a little more aggressive. Is it not a fact that roses are much overplanted and feature too prominently in the majority of gardens? Even in June–July, when at the height of their glory, I think as I look at the endless succession of front gardens that line our roadsides what a horrible jarring jumble of strong colours they comprise. True, they don't have to be a jarring jumble when selected and arranged by someone with an eye for what goes with what. But still, when the flowers have gone, you're left with singularly unlovely shrubs occupying all the most prominent positions.

A garden can, by careful mixed planting, be made to absorb a certain number of non-cooperating shrubs such as these, but not too many. It is a question of getting your proportions right. If we can train ourselves to make our roses take their place among other kinds of shrubs and with herbaceous plants, with lilies and alliums and certain other bulbs, and a few clematis, then we shall enjoy their contribution far more but we shall not be able to grow so many of them.

There must surely be some roses that you're growing now but could do very well without. Better, in fact. What about those muscular modern Hybrid Tea rose bushes, for instance, with their great thick stems, huge thorns and coarse foliage? You must have their blooms for cutting, did I hear you say? But really, you know, they don't look much less stiff or more endearing when you've arranged them in the house. A bunch of stocks or pinks would look far more relaxed and be a great deal freer with their scent.

The Floribundas, then? You haven't the time for annual bedding and they take its place. What a pity. Because actually the notion that roses are labour-saving is a complete fallacy. I don't hold it against any plant that it makes work if it's worth it in other ways, but to pretend that it doesn't make work is another matter. I won't run through the kinds of attention that roses demand, here, but having seen to their needs, beds of Floribundas are very inflexible. There they are, making the same display year after year with only their blossom to recommend them. The bushes themselves are a dreary sight for much of the time. Now, bedding out (as I have indicated in another chapter) gives you endless scope for change and experiment at least twice a year; three times if you're enterprising.

If a permanent planting is *de rigueur*, you could include shapely and harmonious ingredients like junipers and cistuses, santolinas and yuccas (for contrast), that give pleasure for most of the time and among which a few roses would contribute their own special ebullience in their season without hogging it all the time.

Some gardeners, while reviling the HT and Floribunda rose, go all out for the old shrub roses and themselves get bogged down. I grant their appeal; flower colour, shape and scent all work together. Colours may be bright, but there is never the crude admixture of yellow and geranium red in them that we find (a) since Pernet-Ducher introduced the blood of the bright yellow Austrian briar, *Rosa foetida,* into the old roses, and (b) since the post-war mutation which introduced pure red geranium pigment into our roses and all the hot colours that have ensued.

But the old shrub roses are exacting plants. They demand a deal of careful management. I remember being shown a garden devoted to this type of rose of which its owner, clearly dissatisfied, as well she might have been, exclaimed that she was never again going to allow other plants among and under her roses (there were a few spindly nigellas, with an under-the-axe look about them). One has to do things for the roses, she remarked, implying that any competition was a threat to their well-being. And yet there were acres (they felt like acres, anyway) of unappetizing ground among the shrubs, many of which were looking wretched, as though longing for a package holiday on the Costa Brava.

If only, I thought for the umpteenth time, if only roses can be absorbed into our gardening, then the dog can wag its tail once more and a proper balance is restored.

John Treasure manages them beautifully in his garden at Burford House (in the Teme valley on the Shropshire/Worcestershire border). In the first place he knows how to prune and train them, and most need a good deal of this every year. He keeps those that are inclined to make long, leggy shoots down to 3½ ft at flowering time so that they are fully furnished with leaves and blossom right to the ground. A triangle of stout chestnut spiles forms a 3½-ft tall framework round a bush like the Gallicas 'Duc de Guiche' and 'Hippolyte' or the bright pink Damask 'Ispahan'. To this the roses' new canes are trained in a spiral, being tied in individually to a post only as often as is necessary to keep them in position. The posts and string, I may add, are of muted natural colouring, so that you only see them if you're looking for the mechanics of the job.

Pfitzer's junipers, of oblique rather than prostrate habit, are used a good deal among or in front of the roses, and I especially noticed and liked (when on a June visit) the way his 'Duc de Guiche', which has flat, double carmine flowers with an open greenish centre, had one of its canes straying 'naturally' forwards into a juniper, whereas the rest were secured behind and at a higher level to their chestnut supports.

These mixed borders with their accent on old roses are very wide, which gives elbow room for large specimens and allows considerable planting choice. At the back a 10-ft *Rosa glauca* (syn. *rubrifolia*) has a splendidly vigorous *R. × odorata* 'Mutabilis' to one side. It is 6 ft tall and not a bit leggy, and anyway you wouldn't have seen its legs because of a group in front of the 3-ft herbaceous perennial *Baptisia australis*, with indigo blue, lupin-like spikes of clean colouring and outline. 'Duc de Guiche' was by the side of this and then, in front, the juniper with a summer-flowering clematis threading through it. You can cut this type hard back almost to ground level annually, without loss of blossom, and in this way it never becomes an embarrassment to its supporting neighbours.

In front of the Damask rose 'De Resht' – bright carmine, very double and flat – and the deep crimson 'Souvenir d'Alphonse Lavallée', which is of semi-sprawling habit but in need of no support, there were plantings to give later interest of *Iris pallida* 'Argentea Variegata', an *Agapanthus campanulatus* cv., and a pleasing perennial labiate that I once grew from Thompson & Morgan seed, *Scutellaria canescens*. It has 3-ft spikes of soft greyish-blue flowers.

Cistuses are not unlike roses and I liked the juxtaposition of *Cistus × purpureus*, which is rosy magenta, 3½ ft high, with Centifolia rose 'De Meaux', whose tiny, absolutely formal, flat double flowers are only 2 in. across. They are whorled in the centre, thanks to their superabundance of petals. This is a weakish grower, only 3 ft high. With and behind it, in delightful contrast, *Buddleja alternifolia* 'Argentea', so much prettier and a little less vigorous than the type-plant with plain green leaves. Silver shoots and wands of lavender blossom, at its best in June.

Another good grouping combined the white flowers (pink in the bud) of *Cistus* × *hybridus* (syn. *C.* × *corbariensis*) against a white York rose, *Rosa* × *alba,* which has glaucous foliage; a stooled plant of *Eucalyptus gunnii,* bearing its rounded, glaucous, juvenile foliage only, a huge mound behind of the purple-leaved form of *Cotinus coggygria,* the Venetian sumach, and an enviably large specimen of the glaucous *Berberis temolaica* to one side. Only a foreground apron of heathers seemed a bit of a wasted opportunity.

With all their free-and-easy appearance, these roses at Burford House are carefully pruned and trained annually and repeatedly sprayed against fungal diseases in the growing season. The old roses need this attention just as much as modern varieties do.

Another garden where the owners, the Allan Camerons, love roses but let them take their place with other garden flowers is at Allangrange on the Black Isle near Inverness. The mauve panicles of *Campanula lactiflora* look well with every kind of large-growing rose, in this case in front of the soft pink Alba rose 'Céleste'. White Rugosas made a pleasing background to this same campanula, to blue delphiniums and the uncompromising scarlet *Lychnis chalcedonica.* It is always worth finding the right place for this.

White (and grey) delphiniums look good with white roses, there being such a contrast in form. Here it was the vigorous double shrub rose 'Mme Hardy' behind and the modern single-flowered 'White Wings' in front. I'm very fond of 'White Wings', and its purple stamens make a special feature of the flower centre, but alas that it should make such a miserably stalky plant.

The single white *Rosa* 'Paulii' is so overwhelmingly vigorous that it needs a great deal of space. At Allangrange it billowed in front of *Hydrangea anomala* subsp. *petiolaris,* growing up an old tree, and there were white foxgloves all around.

Sissinghurst Castle gardens are especially famed for their roses, but so well are they absorbed into and digested by their surroundings that when someone recently referred to their rose garden, I said, 'Where's that?' It hadn't occurred to me that the roses were thicker on the ground in one place than in another, and there is so much else of interest growing around and near them that you never get that terrible feeling, familiar to old-rose nuts, that after the middle of July the year is over.

There's no part of my garden where roses are excluded either. We do also have a rose garden. It was designed and planted before the First World War and it is hard to think of it in any other role. But the roses are a varied lot and of every age up to sixty years. And I allow other plants to interlope: violets, hyacinths, teazles, and the purple-leaved, yellow-flowered *Oxalis corniculata* var. *atropurpurea*. The tall, stemmy, purple-flowered *Verbena bonariensis* has introduced itself and is a happy ingredient, and around the margins where weedkillers don't reach there is a pink-and-white annual balsam that would take the whole place over given half a chance. Tom Wright, in his book on the gardens of Kent, Sussex and Surrey (in Batsford's *Gardens of Britain* series), kindly described this as 'probably the most successful rose garden of those mentioned in this book'. Any dedicated rosarian would strenuously disagree, and I cannot win prizes from it even in our local flower show. But, within a firm, well-designed framework, it is a happy community.

It is true, as I said at the start, that if you treat the rose as just one among many desirable garden flowers, and if you allow it to take its place with the rest, you won't have room to grow as many as if you made of them a speciality. That should be no loss, considering the wonderful range of plants that beckons to us and insists that they have comparable claims on our favours. And it is certain that our practices in the art of gardening will be greatly enriched when the rose is considered as a companion for other plants rather than as an isolate.

LATE-FLOWERING SHRUBS

Because most shrubs are spring flowerers, there is a premium on those that perform at other seasons. It is well known that roses, hydrangeas, fuchsias, buddleias, heathers, hypericums, hebes and shrubby potentillas, which all start their seasons at midsummer, run on into August and the autumn. I shall give them a miss here so as to make room for less familiar late-flowering candidates. Curiously, it will be found that a great many of them are white-flowered, but that need worry no one.

I've never yet heard anyone complain, for instance, about eucryphias being white. Undoubtedly the showiest of these, having the largest flowers, is *Eucryphia* × *nymansensis* 'Nymansay'. Another advantage it has is that, with only a little guidance from the secateurs, it is easily retained as a narrow, though eventually tall, specimen. Again, it puts up with quite a bit of lime in the soil, which most eucryphias resent. Even so, 'Nymansay' is not my favourite. Dramatic when its blossom, filled with red-tipped stamens, suddenly unfurls in August, it yet quickly shows distress signals when the sun is hot, the flowers becoming limp and jaded. Their season, given a heat wave, is no more than a fortnight. And the leaves, which are all you have to sustain you for the remainder of the year, are coarse. Up to half of them are shed in the autumn . (one of this hybrid's parents being deciduous), so that the other half looks threadbare.

I much prefer another (fully) evergreen hybrid, *E.* × *intermedia*, albeit calcifuge. It grows bushier and bulkier though equally tall, but its leaves and leaflets are smaller and neater. The flowers are smaller, too, but that doesn't seem to matter, if there are plenty of them. Often there are not. 'Rostrevor' is by far the most desirable clone in this respect, but unfortunately a great many different clones, some quite shy-flowering, have had the cv. name 'Rostrevor' applied to them. If you get the true 'Rostrevor', and I am assured by Mr. Charles Puddle that the one they market from his nursery at Bodnant is this, you're on to a really good thing. The flowering season is more extended than that of 'Nymansay' and is usefully centred upon September.

But perhaps the deciduous Chilean species *E. glutinosa* is the best of the lot on acid or neutral soils. The fact of its being naked in winter matters not at all, and its foliage never fails to colour glowingly before shedding.

I appreciate that hybrids generally make hardier, more vigorous and reliable, often showier garden plants than the species that gave rise to them, and yet, having gone through and finally wearied of the hybrids, one comes back to the species for refreshment, as to the sparkling source of a river. The majority of escallonias are serviceable but second-rate shrubs. *Escallonia bifida* (syn. *E. montevidensis*) is in quite another class. It is the last of its genus to flower – not till October after a cool summer but a month earlier if it has been warm. Then it becomes radiant with large, domed panicles of pure white blossom more star-shaped than tubular. Red admirals and tortoiseshells adore it and show up dramatically against their snowy background. This is a vigorous shrub but its growth is soft and brittle, easily damaged by frost and wind. It needs a warm position, preferably near a sunny wall, and it deserves this.

Escallonia 'Iveyi' has *E. bifida* as one parent and it is a very fine evergreen shrub. I have two vast specimens; both came through (with a mauling from which they soon recovered) the 1962–3 winter, so this may be accounted hardy in the Midlands and south or in any coastal area, provided it is sensibly sited with shelter from the coldest winds. Its lustrous green leaves turn yellow before shedding, and this is quite a pleasing feature, most noticeable in spring. Again its flowers are white, but tubular before expanding at the mouth. The panicles are smaller and narrower than in *E. bifida*. Its season is much earlier, in late July, though not starting till August in cooler seasons. It lasts for two or three weeks.

Privets tend to be reviled as a group, and yet there is surprising quality to be found among them. The leaves of *Ligustrum lucidum* are as handsome and evergreen as a camellia's. It grows to a sizeable tree in time, and carries its showy white blossom in September. *L. quihoui* (its name celebrates Monsieur Quihou, onetime superintendent of a Parisian public garden) is a more or less deciduous shrub with small, narrow leaves carried on slender, wand-like shoots. It grows to a fair size, perhaps 10 ft, and should be given a sunny position in the cooler west and north if it is to mature and open its elegant panicles of tiny white privet flowers. They are borne at the extremities of the previous year's young wands. In pruning, then, you remove flowered shoots while leaving the new wands intact. Being a privet this has, of course, the characteristic privet scent, but there's no call to hold handkerchief to nose; you can easily accustom yourself to a smell for the sake of a good plant.

The fuss people make about *Clerodendrum bungei* (erstwhile *C. foetidum*) is quite extraordinary, although you have to bruise its leaves before you get its somewhat pungent but quite inoffensive smell. The flowers have a sweet scent, carried agreeably on the air, and this is another butterfly plant. It is a suckering sub-shrub, coarsely handsome, with bold heart leaves on cane-like shoots. The domed pink inflorescences are packed with dark buds and paler blossom, opening throughout the autumn until frost intervenes. Frost is this plant's chief enemy. If its young shoots are caught in spring, the autumn display will be so long delayed as to catch it at the other end before its display has been completed. Again, if the old stems are frosted to ground level in winter, the young canes are apt to develop too late. But if it starts with some old wood surviving in the spring, you'll have value from it. My colony is now so tall at 12 ft that I scarcely see its skyward-facing flowers, but I smell them every time I pass. *C. bungei* colonizes under trees, flourishing as well in shade as in sun.

So does *Sambucus ebulus*, which has a similar suckering habit though wholly herbaceous. I'm squeezing it in because, being an elder, it must have the soul of a shrub. Its white flowers, often tinged pink, are borne in typical elder corymbs, but terminally on the young canes at 3 or 4 ft. They look remarkably fresh and are shortly followed by black berries. This species, in Kashmir, grows in huge drifts much as bracken does with us. And it is just as wet, after rain, to walk through.

Another good suckerer for autumn flowers is the Japanese angelica tree, *Aralia elata*. You usually see it as a sparse thicket of gaunt, spiny stems, 10 ft tall, crowned with large doubly pinnate deciduous leaves. In autumn this crown produces a huge inflorescence, 2 ft across and about as high, seething with countless tiny white blossoms, particularly effective after a hot summer and if you can look down on them. The ivy family, *Araliaceae*, to which it belongs, is much given to autumn flowering. Thus *Fatsia japonica*, still listed by the seed houses as *Aralia sieboldii*. It carries the largest evergreen leaves of any shrub hardy in our islands, and the compound umbels of white blossom in October and November are as welcome a bonus as they are unexpected.

Fallopia baldschuanica (syn. *Polygonum baldschuanicum*) is so easy, vigorous and common a twining climber as to be despised by the superior gardener. On the other hand it is the darling of impatient gardeners, who plant it for quick cover over an eyesore and then wonder what to do when the climber exceeds its terms of reference, as it quickly does, and flings itself with uncontrolled abandon over all sorts of neighbouring supports that were never intended to support it. Grown up a tree or over the roof of an old outbuilding or ruin, it looks terrific with its foaming white blossom. And I have never seen it better than where it had strayed into a common elder, hung with heavy bunches of purple and black berries.

Pileostegia viburnoides is another autumn-flowering climber with panicles of tiny white blossom, but this is an evergreen and it is self-clinging. Furthermore, it will thrive on north walls as well as on east, south or west. Its leaves are oval, solid and grooved, having red veins and petioles. It must have good soil, well fed and watered, otherwise it just sits and sulks without making any sort of progress.

As most shrubs that flower late do so on their current season's wood, the actual date when flowering starts depends a good deal on how hard you pruned them. In some cases you won't have pruned them at all.

Clerodendrum trichotomum and its variety *fargesii* are just allowed to get on with it, in most cases, and develop into large bushes up to 12 ft high. They may be depended upon to open their corymbs of modest, though pleasing, white flowers in August and September. But what you're always longing for is a display of their shining porcelain blue fruits, each of which is framed by the flower's persistent calyx lobes, which are by now fleshy and bright pink. *C.t.* var. *fargesii* is reputed to be the more reliable in this respect but climate, the site and season all play their part.

The deciduous *Hoheria glabrata* and *H. lyallii* are the hardiest members of their genus, but they are coarse shrubs with large leaves that for some years tend to hide the floral display in July. If you can make a go of the later-flowering evergreen species, they are more rewarding. August is the season for *H. sexstylosa*, when its pearl buds open into countless scented, bee- and butterfly-infested white blossoms. In youth its leaves are quite small and short, but in maturity they become lanceolate, glossy and handsome the year round. I was extremely proud of my specimen, which grew to a height of 25 ft before it was struck down by the coral spot fungus. I had to start again.

This is not such a very tender species, provided it has shelter from cold winds. It grows tall and handsomely in John Treasure's garden at Burford House, in the west midlands, and there are vicious radiation frosts in this valley site.

Less hardy, but still worth attempting near warm walls in the south of our country, are two recent and very beautiful cultivars of *H. populnea*, introductions from New Zealand. 'Alba Variegata' has a clear, clean variegation, its leaves being irregularly white along their margins, blue-green and green in two shades in the centre. Its white stars are at their best in October. 'Foliis Purpureis' has purple undersides to the leaves and also flowers in October. Both received Awards of Merit from the R.H.S. in 1976 and 1977 respectively.

If my *H. sexstylosa* died of coral spot (caused by the fungus *Nectria cinnabarina*) so did the dwarf buckeye, *Aesculus parviflora*, growing next to it, a year or two earlier. I am in no hurry to replace it. At its best this is really good – a large suckering bush covered with slender white horse chestnut candles in August – but it takes up a lot of space in the average garden for a mere fortnight's display.

On its other flank the hoheria has a specimen, now 8 ft high and I doubt if it will grow any taller with me, of *Luma apiculata* (syn. *Myrtus luma*), from Chile. It is slightly less hardy than the well-loved *M. communis* from the Mediterranean, and its white flowers are smaller, but very charming and prolific. They open in succession from July till December. By the autumn, if the weather has cooperated, you have a combination of purple fruits and white blossom. Only when it grows to tree-like proportions, as it does in Cornwall and Ireland, do you enjoy the bonus of its bright cinnamon-coloured trunks, whose colouring is constantly rejuvenated by the peeling of the bark. I find self-sown seedlings of this myrtle even in my garden but in Ireland it naturalizes spiritedly.

Olearias, the New Zealand daisy bushes, cover a long flowering season between them and in some cases their foliage is much more significant than their flowers. My favourite among the autumn group is *Olearia solandri*, which is often mistaken for *Cassinia leptophylla* subsp. *fulvida*, though it has a more graceful habit. It develops quite stiff but spray-like branches with golden stems and undersides to the tiny leaves. The flowers are so insignificant that they would not be worth mentioning were it not for their strong heliotrope scent.

Olearia avicenniifolia makes a far bolder impression in a coarser style, having broad yet pointed leaves, pale underneath. It carries wide and conspicuous corymbs of small white scented flowers in August and September. This is a sturdy shelter belt shrub for windswept coastal gardens and thrives even in the Outer Hebrides. The hybrid *O.* × *haastii* of which it is one parent is far better known and is probably the hardiest of all olearias, but its small tough oval leaves are abysmally dull. So is the shrub's chunky habit. And yet when it is covered with white blossom in August one cannot help relenting. It is quite a spectacle. I should like to see the crimson *Tropaeolum speciosum* growing through and over it.

Itea ilicifolia needs reasonable shelter (though not necessarily a wall). It may get mercilessly cut in a bad winter but soon recovers. Inevitably one compares it with *Garrya elliptica* because it carries drooping racemes of bright green (with a white frosting) flowers, but they are lemon-scented on the air. Furthermore the glossy evergreen leaves are thinner in texture than the garrya's and edged with mock spines. This is an August flowerer and it catches the public's eye.

The best-known and perhaps showiest indigo is *Indigofera heterantha*, till recently *I. gerardiana* and before that *I. dosua*. It is an accommodating member of the pea family and almost entirely hardy. You can treat it in a number of ways. Hard annual pruning in April allows it to be grown as a four-footer in mixed borders. You can even prune it to the ground, each year. Or, given the space, you can leave it unpruned altogether. It may then grow enormous, 10 or 12 ft high and more across, given shelter. Pruning results in your best display of its pinky-mauve, axillary racemes on the young shoots, coming in August. Unpruned, they develop a good month earlier.

I have not for long grown *I. kirilowii*, first seen at Wakehurst Place. It is an excellent low, front row ingredient for mixed borders, behaving like a herbaceous plant in our climate. Its flowers are larger, in more substantial spikelets than the last species, of a clear pinkish shade, while the pinnate leaves are a pleasing foil. It roots easily from cuttings, too; *I. heterantha* less so.

Blue flowers are rare in late summer and autumn, especially blue-flowered shrubs, so one must not pass over caryopteris. The hybrid *C. × clandonensis* produces the showiest garden plants, and 'Arthur Simmonds' is the result of this cross the first time it was made. Still good at its best, but it is extraordinary how blue varies in intensity in different gardens, presumably on account of soil differences. We have richer blue cultivars nowadays in 'Heavenly Blue', 'Kew Blue' and 'Ferndown'. Quite how these differ has yet to be established. As regularly pruned bushes they grow about 4 ft high and contrast strikingly with yellow-flowered shrubby potentillas.

Nandina domestica, whose foliage is often likened to a bamboo's, is rather a pet and not large-growing, as a rule. Evergreen, its foliage is purple when young and it carries modest but pleasing panicles of white blossom in our season. In warmer countries nandinas are capable of setting spectacular crops of red berries, but in Britain they flower too late to ripen them. The leaves appear to be many and small but are actually few and large, being much divided into slender lance-shaped segments.

There are late and early flowering tamarisks; *Tamarix ramosissima* (syn. *pentandra*) is the one for our purpose, and as it flowers on its young shoots it can be pruned to within 2 ft of the ground each winter. Through a mist of feathery green foliage, the pink flowers appear in spikelets over a long season. We could make a lot more of this shrub. I can visualize a large group of it in front of the purple-leaved form of *Pittosporum tenuifolium*, for instance, or queening it above great splashy clumps of pink hortensias, with a blue hibiscus ('Coelestis' or 'Oiseau Bleu') in the wings.

There is a deeper, richer pink clone of this tamarisk called 'Pink Cascade', but it needs a hotter summer than we can be sure of providing, even in the south. In good years it flowers well, towards the end of September, but in bad years not at all.

Erythrina crista-galli is sometimes called the coral tree. It isn't even a bush in our climate, when grown outside, but a 4-ft herbaceous plant dying back in winter to a woody stock. It is leguminous and carries terminal racemes of tremendously exciting pea flowers in which the keel is much expanded, being shaped like a painter's palette. The colour is coral red. It is cv. 'Compacta' that you need, as this is early maturing. Even so, the plant needs as hot a position as you can provide. It will start flowering in August, at the end of a hot summer, but not till October, if then, if it was cool. This is an oft-told tale with autumn-flowering shrubs.

I will conclude with a periwinkle, as periwinkles are always classed as shrubs. They all of them carry the odd flower at any season, but *Vinca difformis* concentrates its main blossoming into autumn and runs on into winter, by which time the flowers are white and only half their original size. In September they are pale mauve, and the shape is intriguing with petal tips cut at an angle, like the blades of a propeller.

ATTENDING TO A SHRUBBERY

February offers us our last chance before the onrush of spring to look about us in a sane and rational frame of mind. One good direction in which to look is towards our shrubs. To illustrate the kind of attention we might give them, I am taking an imaginary instance of a mature shrubbery. It is one-sided, 15 ft deep, and backed by a high hedge or fence. It is (apart from its clematis) stocked entirely with bread-and-butter shrubs such as a wholesale nurseryman will produce with the minimum of trouble for the containerized plant trade at garden centres. In fact I have assembled my shrubs from just such a nursery list and I have arranged them in a workable manner on paper (although in practice I would draw on a wider range of material including herbaceous plants). The border is reasonably sunny; the soil on the heavy side, slightly acid. We start at the left end and will consider the shrubbery contents item by item through the eyes of its owner.

At the back is a large and unruly specimen of *Forsythia × intermedia* 'Spectabilis'. Our gardener realizes now what he didn't at the time he planted it: that there are considerably better forsythias around than this one. He has been meaning from year to year to replace it with the broader-petalled cultivar 'Lynwood Variety', but has every time been put off by the fact that his forsythia is always smothered in flower buds when he has the time to extract it. By the time its flowering is past, in late April, his mind is on other matters. This time, however, he will not be caught out. He picks armfuls of its budded branches, bringing some straight into the warmth of the house and putting others in a bucket of water in a cold shed to be forced into bloom at a later date. Now he feels better about grubbing and replacing the old bush.

In front of it is the indispensable gold-variegated evergreen *Elaeagnus pungens* 'Maculata'. He goes over this with secateurs, removing shoots that have reverted to plain green, which some invariably do each year. In front of it on the border's front left corner he has *Cotoneaster horizontalis*, bracketing forward over the path. Although already breaking into new leaf, it is a trifle dull in spring, and so he plants the Alpina clematis 'White Moth' behind it to train and trail lightly over the cotoneaster's surface. Too vigorous a clematis could be damaging, but this one is of very moderate growth and can anyway be pruned back without detriment to its own performance, immediately after its May flowering season. To the cotoneaster's right are two specimens (one forward, one back) of *Caryopteris* × *clandonensis*. He prunes these back to a permanent framework each late March; it is a little early for safety yet. This is a good clone, but 'Kew Blue' is even richer in colouring, and if his specimens show signs of weakening, which they do not after ten years, then he will replace them with the superior cultivar. Caryopteris look threadbare in the early part of the year and these are underplanted with snowdrops and violets. To the right, at the border's margin, are two plants of *Potentilla fruticosa* 'Elizabeth', once known as 'Arbuscula'. Its pale yellow moon-flowers go well with the caryopteris, though at a lower level, and its season is far longer. He does not make a rule of pruning these, but cuts them hard back into old wood every third or fourth year, and this he does now, as they are perfectly hardy and will come to no harm.

To the potentilla's right and still at the border's front margin is *Deutzia* × *rosea* 'Carminea', a 3-ft tall May flowerer with arching swags of pink blossom. He should have pruned this immediately after its last flowering, but never got round to it, and no harm comes of delaying the operation till winter when you have the added advantage of being able to see the structure of the bush unencumbered by foliage. All flowered shoots (easily discernible by the remnants of old flowers remaining on them) are cut back to where a

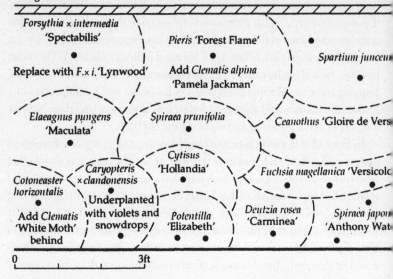

Hedge or Fence

Forsythia × intermedia
'Spectabilis'
●
Replace with F.× i. 'Lynwood'

Pieris 'Forest Flame'
●
Add Clematis alpina
'Pamela Jackman'

Spartium junceu
●

Elaeagnus pungens
'Maculata'

Spiraea prunifolia
●

Ceanothus 'Gloire de Vers
●

Cytisus
'Hollandia'
●

Caryopteris
× clandonensis
●

Cotoneaster
horizontalis
●

Fuchsia magellanica 'Versicolo
●

Underplanted
with violets and
snowdrops

Add Clematis
'White Moth'
behind

Potentilla
'Elizabeth'
●

Deutzia rosea
'Carminea'
●

Spiraea japon
'Anthony Wat
●

0   3ft

young, unbranched, unflowered growth with bark of warm brown colouring arises. If there is none such, he cuts right back to the ground. In pulling out the pruned shoots he finds that several of them have layered themselves near their flexible tips. He cuts these off just behind their roots and pots them individually. They will make nice presents.

Behind the potentillas and flowering at the same time as the deutzia is a bush *of Cytisus* 'Hollandia', a 4-ft broom of dusky pink colouring. He makes sure that its stake and tie are secure and will last another year. Like most brooms it needs support throughout its working life. To set this and the deutzia off in their spring season, the 6-ft *Spiraea prunifolia* is sited behind the broom. It has double white, green-eyed flowers and vivid autumn colouring. He prunes it like the deutzia. And behind this, at the back of the border again, the evergreen *Pieris* 'Forest Flame', which has the double merit of white flower clusters in spring at the same time as brilliant red new foliage. To this, now 9 ft tall, he adds another form of Alpina clematis: 'Pamela Jackman', with deep blue flowers that will contrast marvellously with the red and white of their host.

Moving along the back of the border he comes to a triangular group of *Spartium junceum*, the Spanish broom, and (on its right) a double mock orange, *Philadelphus* 'Virginal'. Our gardener once saw these associated most effectively in a cut-flower arrangement, but it was only after he had put

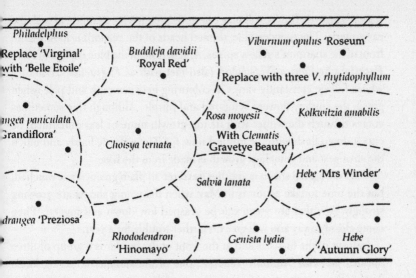

Philadelphus
Replace 'Virginal'
with 'Belle Etoile'

Buddleja davidii
'Royal Red'

Viburnum opulus 'Roseum'

Replace with three V. rhytidophyllum

ngea paniculata
Grandiflora'

Choisya ternata

Rosa moyesii

With Clematis
'Gravetye Beauty'

Kolkwitzia amabilis

Salvia lanata

Hebe 'Mrs Winder'

drangea 'Preziosa'

Rhododendron
'Hinomayo'

Genista lydia

Hebe
'Autumn Glory'

the same combination into practice in his own garden that two vital points
gradually dawned on him. Given the hard annual pruning that the books
told him was appropriate, the broom always flowered too late to coincide
with the philadelphus, so he gave this up, merely tidying off the dead shoot
tips that had developed over the past year. This, however, means that the
brooms have become large and ungainly and need extra-solid staking. They
are obviously nearing the end of their span and will have to be replaced. Not
just yet, however, as he first wants to replace the philadelphus.

'Virginal' is a good old cultivar but not always dependably free-flowering.
Often there is far too high a proportion of leaf to flower. And this was the
other thing he learned: that, in the flower arrangement, the philadelphus
had had all its boring green foliage stripped. This he could not emulate
on the growing plant. However, he decides to replace 'Virginal' with the
very large-flowered single white 'Belle Etoile'. There is a faint purple flush
near the centre of the flower. By the time this replacement has become
established he will replant the broom with youngsters raised from his own
seed sown this spring. Seedlings grow with lightning rapidity.

In front of the spartiums are two *Ceanothus* × *delileanus* 'Gloire de
Versailles', side by side. Ugly shrubs of makeshift construction, they
annually need pruning hard back to within an inch of the base of the last

season's growth. But he will play safe and defer this for a month, as with the caryopteris. The powder-blue, domed heads of the ceanothus look well in front of the spartium's yellow spikes. And in front of the blue is a line of three *Fuchsia magellanica* 'Versicolor' (also met with as *F. gracilis* 'Tricolor'), whose foliage constantly varies in colouring between pink and pale ashen green, the spidery flowers being red and purple. Although this sometimes comes through the winter with its top-growth more or less unharmed, he cuts it down flush with the ground now, for it is perfectly hardy and much the strongest and comeliest growth is made from the base.

This variegated shrub is another 'reverter' to plain green, if not watched, but the time to take action is in May, when the young shoots are growing strongly. Any that are green can be grasped low down and tugged. They come cleanly away and will give no further trouble for a year.

In front of the fuchsias and at the front of the border is a group of three *Spiraea japonica* 'Anthony Waterer', one of the summer-flowering kinds that produces at the tips of young shoots of the current season. It therefore needs more or less hard annual pruning. As he does not want it to compete in height with the fuchsia, he cuts the whole shrub to the ground (like the fuchsia) annually, and again, since it is hardy, he does this now. To the right of the spiraeas is another group of three, this time the hydrangea called 'Preziosa', which flowers successively from July till late autumn, but on side and terminal shoots springing from older wood. So he sorts through these bushes now by thinning out all the oldest, weakest branches, right down to the ground. This entails the annual removal of about a quarter of the bush. Branches that are left are left intact. They are *not* tipped, since this would remove some of the year's strongest flowering shoots. But behind this bun-headed hydrangea is another (three of it again) in total contrast: the white-flowered *H. paniculata* 'Grandiflora', whose inflorescences are conical. This does flower on its young wood and he prunes it now as hard as he will the ceanothus, so as to encourage strong young shoots carrying vulgarly large panicles next September that will make his friends ooh and ah.

To the right and making a firm evergreen unit in the border's centre grows *Choisya ternata*, the Mexican orange. What a valuable shrub; its trifoliate leaves so glossy and brightly toned, its flattened heads of waxy white blossoms scenting the garden in May. After a hot summer it flowers again, even more effectively and with larger individual blooms, in the autumn. But its branches are brittle, and the weight of some heavy, wet snow in January, which he didn't get out to knock off quickly enough, broke two large pieces. He tidies them off with a saw. Choisyas break (meaning 'shoot', this time) splendidly from old wood, so by the end of another growing season the damage will be repaired.

To flower at the same time in front of it he has two plants of the evergreen azalea, 'Hinomayo'. Perhaps evergreen is a slightly optimistic description, as it lost more than half its foliage in the early winter but still looks presentable. Its flowers are pure pink, unsullied with puce, and he loves the plant's habit of throwing up mushroom-like shoots above its main framework, thereby escaping the squat, dumpy appearance of so many dwarf azaleas. So it needs no pruning, but he gives it a special, thick mulching over its roots of leafmould from leaves collected and penned up two autumns ago. These will help keep it moist through any dry spells that may develop during the growing season, for Kurume azaleas have their roots near the surface and can quite easily die in a drought, if not succoured.

Behind the choisya is *Buddleja davidii* 'Royal Red' which, like the caryopteris and ceanothus, he will prune hard back in a month's time. Actually he has half a mind to prune it now and hope that he may kill it. At first he was thrilled by its reddish-purple colouring but has since wearied of its lifeless matt quality. He would prefer the green-and-yellow-leaved sport from this called 'Harlequin', but his garden centre doesn't stock it.

However, he is adamant about ridding himself of *Viburnum opulus* 'Roseum' (syn. *V.o.* 'Sterile'), the commonest of all snowball trees, to the buddleia's right. It is tremendously vigorous – he has already cut it back to a stump on one previous occasion, so as to allow more breathing space. Furthermore its flowering, in May, only lasts a fortnight and there is no pleasure in the shrub for the rest of the year. It is a garden form of the wild guelder rose. The latter has beautiful red clusters of shining fruits in its season, but this sterile fellow has none. He will replace it with a closely planted cluster of three *Viburnum rhytidophyllum* seedlings. One would be plenty to fill the allotted space but would probably not set fruit, which is what he particularly wants. The three will pollinate each other and give him, in August, abundant clusters of red, later turning black. He likes its veined evergreen foliage too, and the fact that the off-white flowers are no great shakes can be overlooked.

To left front he has *Rosa moyesii*, whose deep, wine-red single flowers (a richer colour, these, than are borne by the clone 'Geranium') are followed by large orange hips in August and September. This is good value, but even so the rose is a leggy species and to cover up this ugly part of its anatomy he grows with it a clematis of moderate vigour, one moreover that can be cut to the ground each winter (indeed it often dies back to the ground): 'Gravetye Beauty'. Its deep red flowers expand in late summer and autumn from a slender funnel into a 3½-in wide star. It is somewhat prone to powdery mildew on its leaves and flowers. Twice, therefore, during the growing season, he applies a drench of benomyl in solution to its roots: ¼ oz of the powder (first mixed to a paste in a little water) to 2 gallons of water, applied from a can with a coarse rose to the square yard of soil immediately round the clematis crown. He cuts away all its top growth now; its young basal shoots, purplish-red in colouring, are already clearly visible. Then the rose is pruned, cutting away quite long branches of flowered wood back to strong, unbranched young shoots. This has the effect of eliminating about one third of the bush.

By the rose's side grows *Kolkwitzia amabilis*, a close relation of the weigelas. It should cascade in May and June with pink, funnel-shaped blossoms, yellow in the throat, and it is pruned like the rose and the deutzia just described. His bush has never flowered and he has lately read that some clones of kolkwitzia are very shy. Not so the best-coloured of the lot, 'Pink Cloud', but he can't get this locally and is looking for a supplier. Meanwhile, perhaps his plant will have a change of heart this year, its last chance.

In front of the rose he has a foliage shrub, the grey-leaved *Salix lanata*. It was desperately slow to begin with but is at last making headway. It is a male, he finds, for it wears pretty yellow pussies in the spring. No treatment needed here. To its right, two plants of a hebe grown entirely for its foliage, 'Mrs Winder'. Although our gardener doesn't know this, there are at least four different clones going around that purport to be 'Mrs Winder', so he is lucky to have a good one with a wealth of narrow purple leaves whose colouring is richest in winter. It needs no attention. As much can hardly be claimed for *Hebe* 'Autumn Glory' to right front and on the border's front right corner. This is a marvellous flowerer, with its short indigo spikes, abundant in early summer but always carrying a sprinkle through autumn and the whole of the winter if it stays mild. But it is a sprawler. Every fourth year he has to cut it hard back into old wood and start it off again. He does this now, knowing that the treatment will prevent any flowers from being borne during the coming season.

To its left is a group of *Genista lydia*, whose flowering luckily comes just after the azalea's on its left. The broom is the brightest yellow imaginable, and has a pleasing habit with its sickle-shaped shoots. His three plants have grown into each other and made one unit. He finds it is best not to prune this at all, even though the lowest branches have long since been overlaid and killed by young replacement growth on top. By leaving a dense unit, the ground beneath is utterly prevented from growing weeds. *G. lydia* can be expected to give you six or seven years, after which it must be replaced, but it has no cultural requirements.

When the shrubs have all been attended to individually – those that need it – they receive a top-dressing of whatever organic manure is available. I use deep-litter chicken manure (the litter is sawdust) which has powerful food value, but other possibles include garden compost, peat, leaf mould, ground bark, farmyard manure, spent hops, road sweepings and sewage sludge. Some of these materials are excellent for retaining soil moisture but have no food value. They should be supplemented in March with a balanced fertilizer. John Innes base fertilizer, though expensive, is first rate. The hoof and horn that it includes releases its nitrogen slowly over a season.

Pine bark or peat, applied thickly enough, will suppress weed seeds, but if weeds are likely to be a problem, I use a killer based on simazine. This destroys all seeds as they germinate. Applied in early spring, its effects usually last throughout the growing season so long as the ground remains undisturbed. It should always be applied when the ground is wet. On light soils in Holland, which dry almost as soon as it stops raining, they make sure by putting on the simazine while the rain is actually falling.

3

SEASONS AND SITUATIONS

SMALL GARDENS, LARGE TREES

'Could I just tell you,' a lady wrote to me from near Southampton, 'of a gardening problem which concerns me personally, must concern many other people, and which, as far as I can see, has been totally ignored in gardening literature? This is the problem of the small garden containing large trees.'

My correspondent's garden is only 75 ft by 45 ft but it contains four large sweet chestnuts (circumference about 7 ft) and they 'suffer from Tree Preservation Orders'; as she puts it. 'The only approach ever given to this problem is the defeatist one of covering the ground with something.' But she wanted to know how to make a varied small garden in those conditions because 'if you have a small patch you don't want to have to stare at bergenias, lamiums, periwinkles etc. unrelieved, the year through! I am sure you see the difference in emphasis, I mean, between the "how to make it tidy and bearable" approach and one which sees the problem as a creative challenge.' Hear, hear. I'm with her all the way. Let's see what can be thought up.

First thoughts must clearly go to the trees themselves. I find myself partly on their side, as mature and (still more) senile chestnuts have terrific personality. However, the sickly perfume of their flowers can be a little trying in July, when we most want to sit in the garden, and their tough leaves in autumn are as infuriating as the plane's and just as slow to rot. They must all be gathered up and, if there's not the space or time for rotting them, they'll have to be burnt. The ground will be in sore need of humus and this will have to be applied, not just once and for all, but constantly replenished by top-dressing. Tree roots will gobble up four-fifths of these oblations but the underplantings will receive the rest.

I shall assume that the area is fairly dry, especially in summer, but not desperately so; that you will not feel obliged to go round with a hose regularly but that, in emergency, you would make a gesture. Your greatest successes will largely derive from bulbous plants and quite lowly herbs, so I think you should leave your widest spaces uncluttered with shrubs. Concentrate these in areas on their own, mainly in the distance, near your boundaries, though clumps of some, like *Danaë*, will make foreground features.

We had better think of these shrubs first. As they are visible the year round it is essential that they should never bore you. You may or may not want to climb something up the trees themselves. If their trunks are a really handsome feature, leave them clear. If not, then *Hydrangea anomala* subsp. *petiolaris* should be considered. Quite apart from its June flowering, the pattern and cinnamon colouring of its naked stems in winter is satisfying.

Euphorbia characias subsp. *wulfenii*, with its columns of dark, blue-green foliage, always pleases except in frosty weather when the leaves roll up and the plant appears to have stomach cramp, but that passes. Light green flowers for months in spring, but eventually they fade and you must not forget to tidy the plant up by removing all its flowered stems. *E. amygdaloides* var. *robbiae* is a much recommended ground cover plant, I know, but it is not just doing a job. It really contributes in a positive way at the 2-ft level with rosettes of dark, lustrous foliage. Again, its lime-green flowers in spring are long-lasting, and again it is all-important to remove them and the stems below when they become shabby. So few gardeners get around to this essential and quickly performed task that this spurge has earned an undeservedly bad name. After all, we're talking about a small garden where we should be ready with every small attention.

Danaë racemosa is a super foliage plant, 2–3 ft tall, clump-forming with narrow, shining, bright green leaves on branched stems that are admirable for cutting. Again it implores your small annual attention of completely removing its old stems each spring as the fleshy young shoots push up to replace them. The sprays you so remove will still be in good nick and can be arranged with daffodils indoors.

So dark is *Helleborus foetidus* that it would be glum indeed under the shade of a tree were it not for the beautiful fingered cut of its leaves and the pale green of its bell flowers all through the late winter and spring. They are stained purple at the mouth. *H. argutifolius* (syn. *H. lividus* subsp. *corsicus*) has trifoliate leaves of a bluish green and saucer flowers in the same season. If you can find a good strain of *H. lividus* itself, the foliage and its marbling are even more desirable.

I know that *Fatsia japonica* is recommended by every garden designer for every town or suburban garden, but there's no getting away from the excellence of its great, shining, palmate leaves, and the white blossom, coming in November, always takes us by grateful surprise. This is another shrub that you will or should never pass without darting forward to tidy away decrepit leaves that have fallen off or else got hung up in the fabric.

You probably detest aucubas (spotted laurels), but if you are among the chosen for enlightenment, then revel in one of the really hectically variegated kinds. *Aucuba japonica* 'Crotonifolia', for my money, with its mad glad rash of spots and flecks, but 'Picturata' with its central flash is decisive and arresting, too.

It is all very well to say that the camellia has a good glossy leaf, but many of the japonica types are pretty coarse in their growth, leaf size and shape. The *williamsii* hybrids between *Camellia japonica* and *C. saluenensis* are a lot better in this respect, as well as being more dependably free-flowering in shade in the cooler parts of Britain. The single pink *C.* × *williamsii* 'J.C. Williams'' would be my choice, since I am allowing one camellia only. Or its paler pink parent *C. saluenensis* itself, provided the situation was sheltered. Were I to plump for a *japonica* cultivar after all, it would possibly be the tall, loose-limbed semi-double red 'Adolphe Audusson', or the single white 'Alba Simplex', with its cylinder of golden stamens.

Likewise allowing but one rhododendron, its foliage must give year-round pleasure. *R. yakushimanum* does that and flowers well also, pink in the bud and young flower, then white. Single plants look self-conscious and dumpy for many years so I should have three (from a benign relation, if possible; they're not cheap). More probably I should choose a taller species, one with bright rust-coloured felting on the undersides of its leaves like *R. mallotum* (blood red in flower), *R. bureavii* (white), *R. campanulatum* (variable in colour, sometimes distinctly nasty, so choose carefully) or *R. arboreum*, which has many variants and would grow eventually into the branches of a chestnut and develop beautiful trunks of its own. Or you might prefer the metallic, glaucous (in youth), ultra-smooth and nicely shaped leaf of *R. cinnabarinum* subsp. *xanthocodon* Concatenans Group, with apricot-coloured bells. Its young foliage imparts a delicious spicy fragrance on the air and so does *R. cinnabarinum* itself.

Mahonia japonica is highly shade-tolerant, and almost indispensable on account of its long winter flowering season from early November to April and the lily-of-the-valley scent which it has passed on to none of its hybrids. But I find its leaves a little depressing, whereas in *M. lomariifolia* (which flowers in November – no scent) the leaf is marvellously structured. But this is none too hardy so perhaps we should have a hybrid between the two after all, because their leaves are pretty good and they make a grand floral display in late autumn ('Lionel Fortescue') or early winter ('Charity', 'Buckland'). Be sure to prune any of them back quite hard each spring, as their natural shape is odiously leggy.

A *Choisya ternata* on the boundary, why not? At least if you live in the south where the Mexican orange can be depended upon to flower freely in shade, often twice a year, May and October. Good, shining, well-shaped leaves and waxy white scented flowers that show up from a distance.

One of the sarcococcas for low-level furnishing; their leaves are elegant and evergreen, their insignificant white flowers are borne in the depths of winter but smell exceedingly sweet and are nice to cut. I think I like *Sarcococca hookeriana* var. *digyna* as well as any, but all have their virtues.

Few daphnes are keen on shade. *Daphne mezereum* is a woodlander but not much to look at when out of flower. The evergreen *D. pontica* is the one, with its excellently moulded, shining foliage borne in rosettes and heavenly night-scented green blossom in May. Our own native *D. laureola* is a pretty shrub, close to the last but flowering in February. You might be better suited by a dwarf variety of this from the Pyrenees, *D.l.* subsp. *philippi*, which is altogether neater and never straggles. It slowly spreads by suckering.

Some evergreen herbaceous plants should be given a chance. Bergenias have largely been overplayed but *Bergenia purpurascens* eludes the habitual leatheriness of the clan by its leaves being small and neat. They are a warm purplish colour in winter, while carmine flowers are borne on long stems in May. *B. ciliata* is practically deciduous, which means that its leaves are not leathery at all. In a good strain they grow large and imposing and are covered with soft hairs. Its flowers are clear pink; very pretty, with purple stamens.

Iris foetidissima is a messy plant at close range but not too bad as a foil to the broad-leafs in a general setting. Brilliant orange seeds are its main asset, in autumn, but the variegated form makes a virtue of its foliage and so does variegated *I. japonica*. Slugs and snails must be kept at bay.

Ferns will be a great temptation, but they do like moisture and humus in abundance so it's rather up to the owner's efforts to make them flourish. But on let-alone principles polypodies, *Polypodium vulgare*, will succeed in dry shade and they have a number of beautiful cultivated forms, especially the crested kinds and also the prettily divided fronds of *P. interjectum* 'Cornubiense'.

Different varieties of ferns should not be planted so near as to interfere one with another. *Saxifraga stolonifera* and *S.s.* 'Cuscutiformis' make suitable fillers. Their rosettes of shell-shaped, slightly fleshy evergreen foliage rise only a few inches above the soil and they spread by overground runners. The former's leaves are patterned with pale green along the veins (which radiate palmately from the stalk) and dark green between. In *S.s.* 'Cuscutiformis' the area between the veins is purple, while the undersides of the leaves (which, admittedly, you don't see), the petioles and runners are bright reddish purple. This is the more exciting species and I have no evidence that it is less hardy. Indeed, *S. stolonifera* laboured long under a tender reputation which seems totally unmerited. Both plants flower charmingly at midsummer. Fifteen-inch-tall sprays of white blossom, in which the lower petals are much elongated, gives them a puckish look.

I should tuck in *Omphalodes cappodicica* under some of the shrubs. Its leaves are dismal, but its intense blue forget-me-not flowers are with us for at least two months each spring and the plant will thrive under the adverse conditions of darkness and drought. You need to be selective with the related lungworts. Here my choice would be *Pulmonaria longifolia*, whose leaves are long and shapely, spotted white and remaining in condition from spring to autumn.

Still in the borage family, *Mertensia virginica* is one of those deciduous woodlanders which flowers in April and rests completely throughout the remainder of the year, so it can be grown where it receives a good deal of light in spring, but will be heavily shaded out later on, when it doesn't matter. Its tubular, pale blue flowers hang in bunches on 18-in. stems and are offset by remarkably smooth, fresh foliage.

We're not discussing a piece of woodland but there is the similarity of heavy shade from late May to October; light in plenty from October till May and moisture to go with it. Many herbs (like the last) and bulbous plants all across the temperate world are geared to take advantage of these circumstances and I think we should, without shame, make up our minds to work with the grain rather than against it. That is, to enjoy a concentration of flowers that bloom between autumn and spring rather than the other way round, as in normal gardening.

It may be objected that most people who garden for pleasure would rather forget they had a garden in just that period. Too bad. In that case they'll have to struggle on with summer flowers like Japanese anemones, phloxes, aconitums, rudbeckias, that do tolerate shade but also require abundant water and nourishment if they're not to be a wan, reproachful collection that, far from giving pleasure, merely remind you of how they should be growing. And let me add that there's far less disappointing weather between autumn and spring than in the summer, because expectation is at a low ebb. Every good day is a gift, a bonus. The fact that it's a short day means that we shall savour it the more intensely, just as we savour the winter flowers in our gardens to a degree that reason tells us is out of all proportion. Reason must give way to sentiment on this point. The *Iris unguicularis* (syn. *I. stylosa*) buds that we pick through winter's months touch us more (pride apart) than all the delphiniums that have ever been staged at the Chelsea Flower Show, even supposing we could grow them like that ourselves.

Come off it, Lloyd. Stylosas are not for shady gardens anyway. Cyclamen are. Concentrate, till you have it in masses, on *Cyclamen hederifolium* (new name for *C. neapolitanum*). It is much the easiest species for our climate and it will put up with such dry conditions in summer that you can plant right up to your tree trunks with it. I still have some of this cyclamen flowering under a lilac in December, yet the first blooms (on other, more precocious plants) appear in early August. There is enormous variation in foliage patterns and shapes – as important in this species as its blooms. The pinky mauve flower colour varies greatly in intensity. The pure white albino is as beautiful as any and comes true from seed. Never buy dried tubers of these hardy cyclamen. They have probably been dug up in the wild, which needs discouraging, but in any case the journey and rough handling often kill them.

C. coum will be the next species to establish. It is smaller, less significant, but still a marvellous flowering plant, brilliant magenta and out from December till March, at which season I have seen it in swaggering contrast with the bright yellow cyclamen-shaped flowers of *Narcissus cyclamineus*. The scented *C. repandum* flowers in spring and is greatly to be coveted, but is easily dissatisfied (with me, anyway). It requires more moisture than the others but will certainly flourish in dense shade.

Of the herbaceous hellebores, the deep plummy *Helleborus orientalis* subsp. *abchasicus* Early Purple Group seldom fails to start flowering by Christmas. True Christmas roses, apart from *H. niger* var. *altifolius*, seldom flower till well into January. I wouldn't hold that against them, but they are stiff and ugly plants. Perhaps this is sour grapes on my part; I've never pleased them. But *H. × hybridus*, loosely umbrellaed as *orientalis* hybrids, is for everybody and includes a fantastic range of colours and markings. To get the best from them, remove their last year's leaves before they start flowering, even though still green.

Arum italicum 'Marmoratum' (syn. *A.i.* 'Pictum') sulked with me for several years after I was first given it because I planted it in too dry a spot under my bay tree. Given reasonable moisture it is a winner, throwing up its first handsomely patterned leaves in autumn; thereafter in ever increasing abundance until late spring. Its dense heads of red berries in late summer are quite glamorous too. Not too dark a position for this, if you are to enjoy the strongest foliage patterning.

I should tend to avoid the whole primrose tribe as they are so persistently set about by sparrows, especially in small urban and suburban gardens. Black cotton works but will tempt you to herd your primroses into one concentrated area for convenience; an outpatients department.

Turning to the bulbous plants, colchicums will take a good deal of shade and make a vital autumn contribution. I suppose if only one could be fitted in it should be *Colchicum speciosum* 'Album', with pure white goblets of considerable size and substance. These will show up from afar, whereas the tiny white bells of the autumn snowflake, *Acis autumnalis* (syn. *Leucojum autumnale*), invite close inspection.

Snowdrops we can have in abundance in moister hollows. The more species and varieties you grow, the more you come to appreciate their finer details and differences, whereas those who are strangers to the subject think that all snowdrops are alike, even taking offence if it is suggested that they are not.

Winter aconites should be happy. As I've never succeeded with them I'm not really the one to say, but I have seen them especially well partnered under a greedy copper beech with the mauve of *Crocus tommasinianus*. If you think that crocuses are captious – and mice can make them seem so – start with this species. It seeds itself madly, gladly all over the place and colouring ranges from white through every shade of mauve to 'Whitewell Purple'. For autumn flowering, *C. speciosus* is almost as prolific and it is just about the handsomest species going, as well as being the easiest. Heavily veined flowers give a rich blue impression offset by startling orange stigmas in the flower's centre.

Some bulbs look such a mess after flowering, when their foliage is slowly dying off, that you begin to wonder if they're worth it. Crocuses are not among them, but any of the coarser, grander daffodils should be firmly excluded. Keep to the little ones, like the rush-leaved hoop-petticoat, *Narcissus bulbocodium*, even prettier in its pale yellow variety *citrinus*; *N. cyclamineus* and most of its hybrids absorb well into their surroundings, and the Lent lily, *N. pseudonarcissus*, not forgetting the miniature version of this, *N. asturiensis*.

You see a great many bluebells in small, shady London gardens, but they are always the Spanish kind, *Hyacinthoides hispanica* (usually listed as *Scilla campanulata*), which is bigger and showier than our native species and also includes pink strains as well as the more usual pale blue and white. I like it – quite. But its fat bells are undoubtedly coarse. English bluebells are more shapely and are happy garden plants, seeding themselves generously. Their bulbs go very deep, so you don't want them in the wrong place.

Do grow *Ornithogalum nutans* where you can enjoy it at close quarters, with its vivid green petals. It puts up with arid, rooty soil. Erythroniums, on the other hand, must not be allowed to become too dry, though shade suits them admirably. Their fascinating turk's-cap flowers come in rose-mauve, bright yellow and glistening white in *Erythronium revolutum* 'Album'.

Some hardy orchids are sure to be happy, and I especially recommend any dactylorhiza, as they increase naturally by division of their tubers, so you can soon work up good colonies. That makes up for having paid a lot for your original. Dactylorhizas were separated from *Orchis* on account of their different root formation, so the common spotted *O. maculata* becomes *D. maculata*, or something of the sort, and with its rich lavender spikes and easy temperament this is one to start with. So is *D. foliosa*, with sumptuous, reddish-purple spikes.

Violets, scented or not, offer great scope, from those that are violet, pink or apricot to others bright yellow or white. The white form of *Viola cucullata* makes dense and dazzling clumps that show up from a great distance. The purple-leaved form of *V. riviniana* appears to have the sulks. It is a good plant, but I'm still searching for the ideal companion to act as a foil. Beth Chatto recommends Bowles's golden grass and snowdrops. A late snowdrop like *Galanthus platyphyllus* would certainly be good.

I do think one should find room for foxgloves, so effective under and among trees in early June. The albino or pale apricot kinds will stand out clearest, especially as the light fades. They will bring our season, as I have conceived it, to a close, though I should bridge the gap till cyclamen and colchicums are with us again by establishing a colony of *Anomatheca* (syn. *Lapeirousia*) *laxa*. Its little ixia-like flowers in pale and dark red, carried on 9-in. stems, open through many months, and it seeds – beautiful, gleaming, wine-red seeds – generously. It has a white-flowered variety that is scarcely less appealing.

DRY WALL FREE-FOR-ALL

The most effective way of dealing with slopes in any but the largest landscape gardens is by dry wall terracing. If you are an alpine gardening enthusiast you'll use the cracks between the uncemented stone faces as homes for a variety of specialities: rarities such as lewisias, haberleas, ramondas and encrusted saxifrages.

But suppose (like me) you don't want, in this context, to be bothered with plants that need cherishing but would just like a jolly assemblage that will look after itself, once established, with no more attention than an annual trimming. Such plants should be able to spread in the walls by their own methods, which will be either by seeding or by suckering.

Another slightly different question worth looking at concerns the kind of plants that can suitably be ranged along the top of a retaining wall; boldish plants (shrubs more often than not) which will stand out like brackets or cascade down a wall face and make a strong impression from a distance.

But I'll take the wall face plants first. Two of the best I have discussed as non-stop flowerers (see page 176), and so of the Mexican daisy, *Erigeron karvinskianus*, and of the ivy-leaved toadflax, *Cymbalaria muralis*, I need say no more except that the daisy needs to be established, originally, as a quite small seedling. It quickly develops a tap root and then becomes difficult to move. So it is no good pulling a plant out of someone else's wall or steps (mine, likely as not); this is too ham-fisted a method.

Both the starry-flowered *Campanula poscharskyana* and the bell-flowered *C. portenschlagiana* (*C. muralis*) will move freely between cracks and it would be hard to have too much of them. Both grow as well in shade as in sun but the former, in a sunny position, develops the added attraction of sun-tanned red stems. Its normal flower colouring being a somewhat washed-out mauve, it is worth seeking out a cultivar either of a deeper campanula blue like 'Stella', or the white. Not pure white; it retains a faint glacial suggestion of blue, rather of the same order as the white Persian lilac, *Syringa* × *persica* 'Alba'. In a slightly more civilized position where it would not be ousted by boorish neighbours, the fairies' thimbles of *C. cochlearifolia* (*C. pusilla*), in blue or white, would be appropriate, for this again spreads by rhizomes.

Erinus alpinus is a quite tiny, rosette-forming plant with magenta, mauve or white flowers from May to July. It is surprisingly adept at looking after itself, seeds freely, and is as happy on a shady face as in sun. So is *Ceratostigma plumbaginoides*. Normally grown as ground cover, I have never seen this poor relation of the plumbago more handsomely displayed than in a dry wall, for it has a creeping rootstock. Its dark blue flowers are not borne till the autumn, and then only in the warmer south and at the end of a warm season. Given these conditions they are prolific and combine with crimson and purple autumn foliage tints.

As most of us are well aware, by now, *Euphorbia amygdaloides* var. *robbiae* is a real thug, but invaluable evergreen ground cover in the most difficult of all circumstances: dry shade. As well as being a runner, this spurge can also start new colonies by self-seeding, and it has done this in our walls in several places. I should not otherwise have thought of it in a mural context. But it looks well, and especially when its lime green inflorescences come out in spring.

I don't know how we ever came to lose *Corydalis lutea*, the yellow fumitory, from Dixter, for it is quite an aggressive plant and most appealing each spring when its clumps are covered with spikes of cheerful yellow tubular flowers. I keep meaning to reintroduce it and even grew some in a pot from seed, which only germinated in the second year, but there my initiative evaporated. There is a charming creamy white species, *C. ochroleuca*, also.

Certain ferns are especially successful colonizers of shady wall faces. The maidenhair spleenwort, *Asplenium trichomanes*, makes starfish rosettes that gradually build into larger clumps. The simple, pinnate fronds are shown off by black stems. This is evergreen, whereas the bladder fern, *Cystopteris fragilis*, is deciduous and of the greatest freshness and delicacy with finely dissected fronds that long retain a pale green colouring. These two ferns will flourish in alkaline conditions, so you won't need to worry if some mortar was included when your walls were built. But if they are truly dry, unrendered walls and the soil behind them is acid, you should include the oak and beech ferns, *Gymnocarpium dryopteris* and *Phegopteris connectilis* (syn. *Thelypteris phegopteris*), both deciduous and with slowly creeping, colonizing rootstocks.

Most suitable of all, perhaps, is polypody, *Polypodium vulgare*, which is the most widespread of all our native ferns next to bracken. It is a creeping evergreen and makes curtains of foliage on a vertical face when well established. There are excellent and easily obtainable cultivated forms of polypody that come in apropos for garden purposes.

I was on the point of leaving the wall's face for the horizontal ledge above it when aubrieta and alyssum suddenly made a reproachful assault on my recollection. Both are apt to be castigated as suburban garden plants. It is only their use that is at fault, not the plants themselves. Mustard yellow alyssum is in too strong and blatant a contrast to purple or carmine aubrieta when the two alternate on rock bank or wall face as though bedded out. Let them make large, loose, untrimmed mats and don't bother too much about equipping yourself with the largest and brightest-flowered aubrietas. A homely range of mauves with here and there a bit of pink or purple slipped in will make for relaxation. The pale yellow *Aurinia saxatilis* 'Citrina' will blend with anything, but is not nearly as vigorous a plant as the unrefined species and I would certainly include one good hummock of the latter somewhere on the wall for it is a gladdening spring colour.

Now for the wall top. You'll probably have *Cotoneaster horizontalis* bracketing forwards here if it's anywhere else in your garden because this is just the sort of place that birds will deposit its seeds. Its stiffness is in striking contrast to the more rounded or cascading outlines of helianthemums, candytuft (*Iberis sempervirens*) and other soft or supple shrubs.

Fuchsias are among the best of these when they are fairly vigorous trailing types. They die back to ground level each winter and must be capable of 2 or 3 ft of growth in a season if they are to hang out really effectively. We have a red and purple one that does this to perfection, but, alas, I've never been able to find out its name. So, as it came from my grandmother's Putney garden, years ago, I call it 'Granny's Weeper'. 'Marinka' and 'Lena' might not be strong enough. The red and purple 'Corallina' would do a job but is too leafy and really a bit of a bore. Incidentally, you might think that fuchsias, which are moisture lovers, would find a perch on a sharply drained wall edge too dry, but such positions are, as a rule, astonishingly moisture-retaining.

Rosemaries are naturals in a wall. I have a special affection for the clone introduced from Corsica by Collingwood Ingram, called 'Benenden Blue'.

I have never made a success of *Convolvulus cneorum* on my Wadhurst clay, but given a light soil (especially if it is chalk) and a sunny, protected position, it is as likely to succeed on a dry wall margin as anywhere, and is so good as to be worth fighting for. This is an evergreen, grey-leaved shrub whose blush-white flowers are borne terminally both spring and autumn. It is an unusual-looking plant and quite exceptional alongside the general pattern set by convolvulus. *Bupleurum fruticosum* is unusual in the same way: there are few shrubs in the cow parsley family. This one has evergreen, shining leaves, and the shrub reclines agreeably. Its greeny-yellow parsley flowers appear in late summer.

The everlasting peas are herbaceous, but no less suitable for that. The white form of *Lathyrus latifolius* is wonderfully pure, and I should also like the bright terracotta (but not in the least pinkish) *L. rotundifolius*. How it acquired a reputation for tenderness beats me. It seems to be as hardy as they come, but you may have a job to find seed – the best way of raising it.

Ceanothus thyrsiflorus var. *repens* is exceptionally hardy within its class of none-too-hardy evergreens, and it is a bold plant with masses of pale yet striking blue flower trusses in May. Any of the cistus tribe should do, but some of these trail more than others. Unless the wall was viciously red, I should incline to the soft pink *Cistus* × *skanbergii*. Its flowers are smallish but carried over a long season.

Rosa wichurana is completely trailing and flowers at a usefully late season, July to September, producing trusses of smallish single white blooms with yellow stamens. The leaves are glossy and almost evergreen but became so martyred to mildew in my garden that I gave up. However, you never know with mildew. You can see that 'Max Graf' derives from this species by its gloss and its prostrate habit, but it has *R. rugosa* as its other parent and no rose is healthier than that. 'Max Graf has the one crop of light magenta single flowers of no special distinction, but it is a good plant.

I have tried the miniature trailer 'Nozomi' but was not long in rejecting it with contumely. It is a ground uncover plant and as you remove the weeds from among its trails it grabs at you with vicious little hooks. Its flowers, produced in one crop, are described as pearl pink but are such a wan shade as almost exactly to match the colour of parched grey soil. So you hardly see them in the normal way. If it had sufficient vigour to cascade forwards over a deep red brick wall face that might look good. It hasn't the vigour.

You need to remember that the plants you wish to cascade over a wall are also likely to grow in the opposite direction, covering the ground, so the better they are at this job the happier you will be. *Genista lydia* is an excellent ground-covering broom if you don't tidy-mindedly remove its dead branches. This broom's sickle-shaped shoots arch forwards over a ledge most becomingly. *Cytisus × kewensis* is not so vigorous and would be an excellent choice where the scale of its surroundings was smaller. Its very pale yellow flowers open in May whereas *G. lydia* comes in a month later with brilliant chrome blossom.

Junipers make a restful contrast to all this colour, and my greatest success has been with *Juniperus sabina* 'Tamariscifolia'. This plant does not grow back from its wall. Its energies are concentrated on cascading down to the next level and its branches are beautifully textured, each shaped to a point, so that it is incapable of looking amorphous or lumpish.

The non-climbing trailers among clematis are suitable. Where too much vigour was not required I should choose the rich-indigo-blue-flowered 'Durandii', but where masses of growth and thousands of blossoms were in order, then *Clematis* 'Praecox' (syn. *C. × jouiniana* 'Praecox'). This is on the blue side of white and blooms with fantastic freedom from July to October, if you plant the early flowering cultivar. A foam of the scented white blossom of *C. flammula* at this season would be enchanting.

In conclusion I must bring in the red valerian, *Centranthus ruber*, which is a wall or cliff plant *par excellence*, though its roots are thick, strong and destructive. Plant it on a wall top and you'll soon have it on the face and at the bottom as well, though you can largely prevent self-seeding by cutting plants hard back to the woody stump immediately after their first, June, flowering. They will then flower again in the autumn; pink, red or white in domed heads that are stylishly presented.

NO MUTUAL ANTAGONISM

'At home,' the friend who edited my *Well-Tempered Garden* once wrote me, 'I am assiduously easing myself up to the higher horticultural spheres of progressively less gay plants. It's rather sad that one's gardening evolution should necessarily be towards the silver, the green, the glaucous and the grey. Like human life, really.' Anyway, sundry euphorbias (*wulfenii*, *robbiae*, *polychroma* and a red one I don't know the name of but which is almost colourful) are all growing away, as is my first hosta, *undulata*. My wife is restrained about them and the children snooty, but they have all got to grow up.' Soon afterwards he emigrated to France where, I imagine, the social pressures around him must have tipped the scales heavily back towards a nineteenth-century style of bedding.

Well, but what about this? Are we of the form, texture and delicate colouring school really such prigs? There is a danger. It is a reaction to the other extreme that we see in so many front gardens: the colour addicts who lurch incontinently from lumps of forsythia and double pink 'Kanzan' cherry, through a blaze of rhododendrons and dumpy blobs of azalea, Floribunda roses, scarlet salvias and so to a dying exit with mop-headed chrysanthemums.

That sort of gardening is a bore, especially as there is so much of it, but at least it is full-bloodedly boring. It wallows in its vulgarity with a sense of enjoyment. The trouble with the form-and-texture-and-colour-harmonies gardener is his (often her) insufferable self-consciousness.

I'd rather choose something from both styles. Plants of good form and with beautiful foliage are sustaining. They keep us interesting company for many months, if not throughout the year. But colour, bright colour, gives us moments of great excitement and all the greater if the display is fleeting. Even the 'Kanzan' cherry, if it is a full-grown specimen that has at last emerged from its gawky youth, is a thrilling sight when only half its buds have expanded. The condition lasts for three days. In a week the tree looks tired and overblown. Oriental poppies are fleeting in the same way, but what style, what exuberance. Their very fragility is a part of the attraction, for it is through our strong sense of mortality that ephemeral beauty is so keenly savoured.

There are certain plants, then, that I do not want to be restrained. If I could grow bougainvillea in my Sussex garden, I should eschew the more civilized and muted red, orange, pink and apricot shades and settle for the original startling magenta. What could look more appropriate against a whitewashed wall? Such flowers as these distil the very essence of light and sunshine, growth, vigour and fulfilment.

Neither should one be afraid of experimenting with one bright colour clashing against another that doesn't complement but shouts at it. All depends on the setting. Nobody (except the farmer) complains about a meadow mixture of red clover (which is really a deep shade of pink with quite a dash of mauve in it) and yellow buttercups. 'Charming,' we say, if we notice it at all; because there is an abundance of green grass as background.

I liked an early colour effect I had in my Long Border, though it made some visitors shudder: the very pure, bright red Lily-flowered tulip

'Dyanito' grouped behind *Bergenia* 'Ballawley', which is a good strong magenta (much cleaner than *B. cordifolia*). Again, the rest of this part of the border was full of foliage; nothing else.

"Geraniums" (pelargoniums, correctly) are well known to clash well, and especially the ivy-leaved kinds. I was intrigued by a model garden exhibit at the Chelsea Flower Show one year. A large collection of ivy-leaveds – red, bright pink, magenta and bright mauve – had been assembled and they looked fine with their own lustrous green scalloped foliage to set them off. Behind this group was another of hybrid rhododendrons having roughly the same flower colourings. They were trying to emulate their partners but signally failed. Not only were their colours a shade dimmer in every case but they were hopelessly handicapped by their dark, cumbersome leaves.

An even more startling essay in colour contrasts was to be seen at Inverewe gardens in north-west Scotland, when I visited them one June. At the back of a herbaceous border a huge patch of *Geranium psilostemon*. This is a 4-ft plant with hosts of rich magenta cranesbill flowers, each with a large black centre. In front of this, *Achillea filipendulina* 'Coronation Gold', which could scarcely be a brighter yellow, and in front of that again, at the border's margin, the panicles of *Astilbe* 'Fanal'. This is pure, deep red. One gasped and then had to admit that, yes, it worked. But all the plants adjacent to these hot groups were of a soothing nature; the spires of purple monkshoods, the domes of *Campanula lactiflora* and hummocks of grey santolina. Slabs of vivid colour with no company other than more slabs of colour are exhausting, indigestible. But in each case that I have quoted the setting was all-important.

Foliage has a poetry of its own. Flowers of a strong design like the subdued mauve globes *of Allium cristophii*, the hooded purple and white of acanthus spikes, the many-branching candelabrums, never brighter than dove mauve, of *Eryngium pandanifolium*, the sturdy lime green columns of *Euphorbia characias* subsp. *wulfenii* – these tie in naturally with the best kind of foliage, often their own. It would be perfectly feasible to garden happily and creatively with such plants and never to introduce a strong flower colour. I could do it myself but in the long run the restriction, even though self-imposed, would be stultifying. Our lives are bound by enough restrictions without wilfully imposing more. The greater the range of plant materials that our gardening can assimilate without becoming an ill-assorted mess, the greater the satisfaction our hobby will afford us.

Bulbs, bedding plants, shrubs and hardy perennials will all find a place in this section, whose purpose is to think out ways to use and combine plants that deserve growing anyway, so as to make the most of them.

If you don't think upon these matters but simply plant for the sake of the plant's good health, you'll come some nasty croppers. The rhododendron tribe seems to bring out the worst in its admirers in this respect. Visit any large woodland garden you care to name between April and June – the Savill Gardens, the Hillier Arboretum, Inverewe gardens, Exbury (I haven't been there and am damning it at hazard) – you name it, they've done it; appalling juxtapositions that make you moan in agony, 'Oh no! they can't do it; it isn't true.' But of course it is true and much admired by the public so what am I carrying on about anyway? Well, I just don't happen to believe that 'it's all a matter of taste' excuses or justifies every indiscriminate use of colour in gardening. I believe there are absolute standards of what goes and what doesn't in matters of taste.

I also realize that I perpetrate crimes against colour in my own gardening; usually by accident and at least I try and do something about them most of the time or if I don't I am apologetic. It was really a mistake my ever introducing *Rhododendron* 'Tessa'. It isn't a big and blowsy hybrid at all but – well, anyway, you get it and live with it for five years and see. It is quite dwarf with small leaves and its flowers open at a most welcome season, in February. If the weather is inclement you can pick it when there's colour in the buds and enjoy them indoors. If your plant gets scraggy because you've allowed other plants to encroach on it, you can cut it all back into old bare wood and it'll break freely.

It's a strong pinky mauve and I planted it behind a carpet of the *brilliant* orange *Crocus flavus* (syn. *C. aureus*). Disaster. I'd forgotten the crocuses were there when I placed the rhododendron. I took a photograph of them – very colourful, I must say, for the time of the year and it's been published in the *Observer* magazine – and then I dug the crocuses out and spread them around in rough grass, which they stud like jewels. If I get around to it I can still have a carpet of crocuses in front of 'Tessa', choosing perhaps *C. sieberi* 'Violet Queen' or *C. chrysanthus* hybrids 'Blue Pearl' or 'Ladykiller' (purple and white) or 'Cream Beauty'. There's no shortage of suitable alternatives. However, seedlings of *C. flavus* are still coming up, so I must bide my time.

In discussing spring meetings I shall not long linger among those that are so grand in scale that we can only hope to enjoy them as gaping visitors. But I did see one beauty in the Savill Gardens at Windsor, on a fickle April day of snow showers and dazzling sunshine against blue-black clouds.. A weeping willow, on which every raining wand was picked out in yellow with only a touch of green, side by side with an Asiatic tree magnolia, *M. campbellii* subsp. *mollicomata*, covered on naked branches with pale pink waterlily blossoms that had largely escaped being frosted.

Another thought-provoking contrast on the same day and in the same garden was where a ditch had been planted with and enthusiastically colonized by the yellow American aroid, *Lysichiton americanus*. Its substantial vertical spathes contrasted with the globular knob heads of *Primula denticulata* on the banks above: especially satisfying because the greater proportion of the latter were the pure white form (the bright pink of *P. rosea* near by did not fit in). The mauves were there too, in a patch of their own, and they worked in well, but yellow and white is a particularly agreeable blending of flower colours. I love doing large flower arrangements in early summer with yellow flag and Dutch irises, branches of *Piptanthus nepalensis* (syn. *P. laburnifolius*) with yellow pea flowers and the white water dropwort, *Oenanthe crocata*, a substantial member of the parsley tribe that grows around our horse pond. It lasts much better in water than cow parsley and hasn't the sickly scent. A yellow-and-white border would be great fun to do.

Of course, it was the form as well as the colour of the arums and primulas that set each other off, the more so as neither was diluted with foliage. That develops after flowering.

I have been for some time meaning to make a match in my own garden, in a shady border, between the white strain of biennial honesty (*Lunaria annua*) and another splendid biennial, *Smyrnium perfoliatum*. This again is of the parsley tribe, 3 ft tall with broad, clasping stem leaves and bracts of a vivid, acid yellow-green. It is a plant that I often see written about, but it seems to be spread around between friends and has seldom entered the seed lists.

You start a colony by scattering seed and then you have to wait three years, most likely, for the seedlings develop slowly under competitive garden conditions. Once established it is self-perpetuating. So is honesty and this white one is a true albino. Its flowers are sweetly night-scented.

Alas, house sparrows have acquired the habit of stripping mine of buds, just as they are about to open, so it is no good making honesty plans while this situation lasts.

The best euphorbia displays are in spring, and one good meeting I arranged, though not of great duration, was of the brilliant, hummocky yellow-green *Euphorbia polychroma* interplanted with white tulips. Whenever possible I like to be able to leave my tulips down, and on my clay soil, which suits them, this usually works well, but the euphorbias became so rooty and aggressive that the tulips were starved out. A better arrangement is *Leucojum aestivum* (and you might as well, in a border setting, grow the larger-flowered 'Gravetye Giant' clone) in front of *Euphorbia palustris*. I cannot think of a spurge I love more than this. Its freshness is a revelation. The flowers are lime green and the inflorescence is fairly substantial, 3 ft tall at flowering but later the plant reaches 5 or even 6 ft and takes up a lot of space. If it is happy, that is, and it does relish a moist heavy soil. It will tolerate full shade but you'll do better to place it where it is caught by sunlight because the flowers then glitter as though they held tiny light-reflecting crystals within them.

The snowflake's green-tipped white bells make a good match, and the great advantage of an early-maturing plant in this context is that a month after flowering it has no further need of light and space, having completed its growing for the season; so the fact of the spurge requiring so much elbow room in the summer is of no consequence.

With its burnt orange flowers, *Euphorbia griffithii* is a dramatically showy plant. Like *E. palustris* it does most of its growing after its April–May flowering, but unlike it this species is a runner. It makes colonies by suckering and generally needs restraining around its perimeter with a spade, once a year. There are several clones about and the best known is 'Fireglow'. This is less invasive than some and its flowers are the brightest I have seen, but its leaves are plain green and a little dull. I have a clone I call 'Dixter' which was originally a seedling from that great plantswoman who lived near me, Hilda Davenport-Jones. This is quite a wanderer, but its merit is in having leaves and stems that are all richly suffused with bronze. The flower colour is very good (as long as you grow it in sun and that is always essential, in this species) and the season a week or two earlier than 'Fireglow'.

'Fireglow' I have as an apron in front of the rich, deep mauve-flowered *Abutilon × suntense*, which I keep in order and prevent from becoming too large and stemmy by cutting it back by a couple of feet all over immediately after flowering, in mid-June. If you grew this abutilon in the open instead of in front of a wall as I do, the euphorbia could make a carpet underneath its sparse branches.

Clone 'Dixter' grows next to a shrub whose flower buds are a similar deep orange shade, so the two form a colour harmony rather than a contrast. This shrub is the inestimable *Ozothamnthus ledifolius*, a Tasmanian composite and yet absolutely hardy. It is very bushy, slowly growing to 4 ft, and I sometimes think what a good low hedging plant it might make as an alternative to box, if you delayed its clipping till June, after its flowering time. It is a thoroughly cheerful evergreen with short, narrow leaves that are dark rich green above but yellow underneath and the tips of all its shoots are yellow where the young leaves are held upright in rosettes. The clustered flower buds are numerous and they keep through most of May this extraordinarily mellow brick shade, eventually becoming frosted, as it were, as the white flower emerges, and soon the inflorescence is all white and less impressive than it was. The plant smells strongly, at all seasons, of stewed prunes.

Omphalodes cappadocica is one of the most enjoyable and adaptable of spring-flowering perennials, with its deep clear blue flowers (related to myosotis and cynoglossum) on foot-tall plants over a long season. It can be treated as a mobile bedding plant, and in this role I had it one year interplanted in equal numbers with the double yellow perennial wallflower, *Erysimum* 'Harpur Crewe', which I'd raised from cuttings taken a year earlier. Although I say it, this worked excellently.

In a static and permanent role I've combined the omphalodes with two other yellow-flowered associates, *Ranunculus gramineus* and the jonquil 'Tittle-tattle'. The ranunculus is a foot-tall, clumpy buttercup with lance leaves of a greyish tone. I pointed this feature out one day to an enthusiastic but botanically ignorant visitor and he asked what a buttercup's leaf normally looked like. I didn't have to seek far to find one and he was struck by its charm. Of course, I reflected, if our buttercups normally went around with grass-like leaves, we should be overjoyed to find one in which they were palmate and finely dissected. It seems almost impossible not to be swayed by rarity as a value.

Ranunculus gramineus has tuberous roots that are easily split up and I should like one year when I'd worked up enough stock to use it as a carpet in bedding out for one of the blue-flowered Dutch irises. *Narcissus* 'Tittle-tattle' has the warm scent that one expects of a jonquil, and it is nearly as late-flowering as the pheasant's eyes. But of a far neater and more compact habit and barely 2 ft tall.

The site in which *O. cappadocica* attracts most attention in my garden is underneath and in front of a double red camellia. The omphalodes is perfectly happy in deep shade, and few plants flower as freely in such a position. But you do want to be sure of starting off with a well coloured strain. Some are a much clearer blue than others.

My first plant was given to me by Collingwood Ingram – 'Cherry' as he was known to all his friends, and his book on *Ornamental Cherries* is a classic on its subject. He had a marvellous eye for a good plant or a good form of a plant, and as his travels took him to many parts of the world he collected and introduced beautiful things. The sort of variations within species that the botanists, who have done so much of our plant hunting, wouldn't take note of, simply because they are not gardeners and have little concept of what, horticulturally, constitutes the good garden plant.

Which brings me to *Rosmarinus officinalis* var. *angustissimus* 'Benenden Blue', which is a narrow-leaved form of the common rosemary which Cherry collected in Corsica. It has deep (not the usual wishy-washy) blue flowers and has proved a first-rate garden plant. Perhaps a trace less hardy than the hardiest, but most of those gardeners who can grow this Mediterranean shrub at all should be safe enough with 'Benenden Blue' if they plant it out in the spring.

It flowers in May and sprawls like any other rosemary but I like that. One can grow things in between its semi-recumbent branches, and I have an excellent strain that Margery Fish gave me of avens, *Geum rivale*, with nodding pink flowers in a soft, warm shade. Next to these two is the dazzling *Prunus glandulosa* 'Alba Plena', a bush plum whose young, wand-like shoots are lined along their entire length (2 to 3 ft) with double white bobbles. Immediately after flowering the whole shrub is cut back to about 9 in. and is thus encouraged to make lots of new wands for the next season. And I might as well mention two other points of interest about the shrub while I'm on it. First, to its detriment, it is apt to be attacked by a fungus disease when just approaching full flower, so that the whole display turns prematurely brown. That is if the weather is at all wet, for the double flowers hold moisture. Second, it gives a fleeting but delightful sub-display in the autumn, when the elegant leaves take on rosy tints.

Beyond the plum I have clumps of golden marjoram, *Origanum vulgare* 'Aureum', whose young foliage is at its brightest at just this moment when the others are in flower.

And across the path I sometimes – but this varies, because they are bedding-out corners where a cross-path meets the main one – have the pale, pale yellow egg-shaped tulip 'Niphetos' (classed as a Darwin but not at all typical) with an interplanting of purple-leaved fennel, whose feathery foliage is its deepest colour at this moment. I got this idea from seeing it used by Brian Halliwell at Kew, and it certainly bears repetition. If you have a stock plant of the fennel somewhere and let it seed, the self-sown seedlings (or you can line out a row of them in July) will be just right for bedding by the autumn.

Where I have a path to line with bedding-out plants, as it might be in front of a lavender hedge, set well back, there's nothing prettier in the spring than a double row of Pomponette Series daisies (*Bellis perennis*). They have neat flowers and quilled petals and are much more stylish than the overblown Monstrosum types. It is usually sold in a mixture, but you can sometimes find separate colour strains in red, pink or white. Any of these are far more effective on their own. I usually get the red, which is a deep crimson shade, and it looks striking when interplanted with *Muscari armeniacum* 'Blue Spike', which is a showy grape hyacinth. If I'm picking a small posy I mix the red daisy with the blue trumpets of *Gentiana acaulis*, but it is difficult to organize this particular spring meeting in the garden itself.

You can never allow these daisies to remain in the garden after their flowering period, even though they are perennials, because they self-sow so abundantly and the seedlings will not so gradually revert to ordinary wild daisies. In fact it is dangerous ever to plant them next to a lawn (unless a flowering lawn is what you want), as the early blooms will have seeded themselves into it before the later ones have completed their spring display.

With pansies and violas it's different. They never become troublesome, even though the self-sown plants do have a tendency to deteriorate and you should hard-heartedly pull out any that are markedly inferior to the original strains. But I always like to leave pansy seedlings until they've shown their faces. Once well established they develop the climbing and interweaving habit that is one of their great charms.

For instance, a strain I once had from Thompson & Morgan called 'Lord Nelson'. I think they listed it under *Viola*, but the differences between florists' pansies and violas have long since blurred. Violas are generally supposed to have plain faces, and these are the kinds I prefer in a garden setting as they make a more definite impression than those that are marked. 'Lord Nelson' is velvety purple, and it looked particularly pretty near the young developing foliage of the hardy *Fuchsia* 'Genii' (cut to the ground annually), this being bright yellow with red stems. And a little further from the border's margin the viola climbed into the branches of *Hebe* 'Glaucophylla Variegata', which has tiny cream-margined evergreen foliage.

EXPERIMENTS IN BEDDING

I love bedding out because of the frequently repeated opportunities it gives me of trying out different plants, new textures and, above all, new colour schemes. All too often, bedding out is treated as a stereotyped ritual, repeating itself over and over again without a flicker of interest or inspiration. To say this is to say that a lot of uninspired people are doing gardening, and this is most obvious in the public sector, which, because it is public, we see most of and where there is a good deal of money available but not, alas, enough to buy talent. All this seems a great waste.

What I propose to do in this section is to go over – in some detail, at times – my experiments and experiences with bedding over a decade in one location. This is an important bed right in front of our house at Great Dixter. I have had so much interest in gardening this piece that I must hope that the interest is communicable, so here goes.

The area is right angular, having a lawn in front (which often gets killed at the margins, as I believe in allowing plants to spill forward on to it during the summer) and two walls at the back – one 5 ft high, the other only 3 ft. The front 4 ft of the area with these boundaries is used for bedding. There's another 5 ft of effective planting space behind this, which used to be occupied for many years with a double row of June-flowering lactiflora peony hybrids. They became more and more diseased, so I at first tried to mitigate their derelict appearance from July onwards by planting a row – which was a hedge from midsummer onwards – of *Fuchsia* 'Mrs Popple' in front of them, losing to these some of my bedding-out area. The fuchsias were cut down annually in winter, but on a very good piece of ground they grew 4 ft tall each year. This in turn proved to be somewhat unsatisfactory.

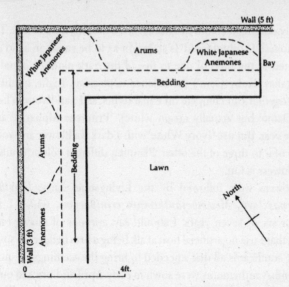

Wall (5 ft)

White Japanese Anemones

Arums

White Japanese Anemones

Bay

Bedding

Arums

Bedding

Lawn

North

Anemones

Wall (3 ft)

0   4ft.

'Mrs Popple' makes a heavy, dark, dull-looking plant, usually with too much foliage and too little blossom in evidence – unless you spray against capsid bugs, which I cannot be bothered to do.

So in 1970 'Mrs Popple' was scrapped and a row of white Japanese anemones substituted; the simple, single white anemone with a green eye surrounded by a circle of yellow stamens. It flowers from early August to mid October. But then it looked wrong to see this beautiful plant regimented into a row. So, two years later, the peonies were finally scrapped and the anemones rearranged in three large informal groups, as shown in the plan, with two groups between them of the white arum lily, *Zantedeschia aethiopica*; this flowers in June and July.

My tale starts in 1969, which was a very good summer for annuals. First, however, I had wallflowers: large alternating blocks of 'Carmine King' and 'Primrose Monarch'. A brown colour is what we probably associate in our minds with a traditional, old style wallflower, but this is not nearly as effective as some of the lighter shades, and effect is what I'm looking for here, as the bed is mostly seen from paths and house windows some distance away.

In the last ten years the mixed wallflower strain called Persian Carpet has taken almost complete command of the amateur gardening market, and this is a pity as mixtures are not nearly as effective (that word again) as straight colours either on their own or in adjacent groups or, sometimes, with dots of a highly dominant colour in a background of some quieter hue. Thus, the old-fashioned 'Cloth of Gold' is so bright as to be garish in solid quantity, but planted in the ratio of one to five of the dark plum-coloured 'Purple Queen' (alias 'Ruby Gem') the balance is pleasant. Light, bright colours planted together can compete on equal terms, as I had 'Ivory White' (alias 'White Dame' but actually cream white), 'Primrose Monarch' and 'Fire King' one year. But use 'Ivory White' with a dark wallflower and you'll want only one of it to three of the other. Planning different colour combinations for wallflowers is fun.

Wallflowers were followed by the Livingstone daisy, *Dorotheanthus bellidiformis* (syn. *Mesembryanthemum criniflorum*), which I had not grown for six or seven years. I should say, apropos of raising half-hardy bedders, that I use no artificial heat at all, being a firm believer in sowing late when the sun heat is all that's needed to bring the seedlings on quickly. So the mesembryanthemums were sown in early April and pricked out 8 by 5. Even though they are succulents it is a great mistake to think you'll get value from them on a starvation diet, which is the story told by so many experts who should know better. The more generously you grow them (within reason) the more generously they'll repay you.

My plants averaged a diameter of 9 in. (which makes for economy in numbers) and were in the usual range of dazzling colours. On a sunny day the flowers expand around 9.30 or 10 a.m. and they remain open till about 5 o'clock in the afternoon, which is pretty good measure as these light-sensitive flowers go. The plants were at their productive peak from late June for a month. You can't lengthen it by dead-heading them. In this case it makes no difference at all. There's nothing for it, you must be ready with a third idea to finish the season off.

I followed with a mixed strain of China asters (*Callistephus*) called Bouquet Powderpuffs, sowing them on 10 June and potting the seedlings off individually into 3½-in. plastics. Apart from not letting them starve, the most important consideration when growing asters, and one that's easily overlooked, is to prevent their growth being distorted and crippled by aphids. It is a green aphid that attacks them so you have to be watchful and spray betimes.

They were planted out at the end of July and for the next two months just remained green. I thought they'd never come to the boil. But they did, flowering all through October, and I was extremely lucky with the weather that autumn. The display was completed unspoilt. But this was a disappointing strain, the plants being stiff, the shoots all so upright as to resemble a swept up broom. The flowers were anemone-centred but in an uninspiring colour range. I've yet to find a strain I like as well as the Princess Series.

I had wallflowers again the next spring, but what they were followed with my records don't tell me. I evidently intended to grow Unwin's hyacinth-flowered antirrhinums but I sowed them too late: mid-September instead of mid-August. When this happens I fill up with a mixture of any bedding plants I can lay my hands on.

In the autumn of 1970 (17 October) I planted up with a carpet of *Erysimum linifolium* and an interplanting of 250 tulip 'Dillenburg'. The erysimum is a wallflower and is sometimes listed under *Cheiranthus*. It is quite a small plant, 9 in. or a foot high, and carries mauve flowers. The shade and intensity of mauve varies a good deal from plant to plant, and if anyone could be bothered to do a spot of selection a very good intense shading could be achieved, eliminating the washier end of the range.

This is a quick developer and even though seed was not sown (in the open) till late June, the plants were already flowering when bedded in October. In the spring, however, their season is later than the common wallflower's and coincides with Siberian types. This was my reason for choosing 'Dillenburg' as a partner. It is the latest-flowering tulip listed. Also it is a pleasing shade of deep orange that looks well against mauve.

That was a success but again there is a veil across what followed in summer '71. No more veils, I promise. That autumn I planted up with sweet Williams in a double mixed strain from Butcher's. I've never grown doubles before and they are charming.

If you sow sweet Williams too late, many of their shoots remain barren and green when they should be flowering. I sow in early April, prick them out and then line them out, and they make magnificent plants without fail always provided I remember to spray them through the late summer and early autumn to prevent their foliage being attacked by carnation rust. This is quite simple, really, because they are given the same copper fungicide as we apply at that season to our potatoes and outdoor tomatoes to stop them catching blight.

The sweet Williams were interplanted with tulips (the same 'Dillenburg'), so that there was interest in spring as well as in high summer. The former were scrapped and the bulbs harvested in late July, being replaced with Clear Crystals Series pansies.

Now there is something very odd about violas and pansies as bedders, and it is this: that no book or article or catalogue in which you read about them ever tells you just what latitude and scope you have in their treatment, or which types should or may be treated in one way, which in another. They are extraordinary plants because they can be grown as biennials for a spring display or as annuals to flower in summer. All one knows about them for certain is that they soon pack up if the weather is too hot and dry. On the other hand if it's too wet, at least on my stiff soil, they go mouldy. They have a wonderful potential but I've had to learn practically everything about them for myself, with the notable exception of how to raise them for spring bedding, which is very well described by Mr Colegrave in his seed catalogue for wholesale growers.

Here, however, I'm concerned with pansies as a late summer and autumn follow-on to sweet Williams. They were in two strains, red and mauve, which I mixed when planting out. Sown in early May, they were lined out and then planted up on the last day of July. They flowered non-stop for the next three months. Elsewhere I left them to overwinter and they flowered again in spring.

But in my main bed, inexorably, the pansies had to make way for wallflowers that were waiting. This time it was a double mixed strain. I've never seen double wallflowers as a seed strain before or since. They certainly flower for much longer than the singles and, at their best, the flowers are prettily shaped in neat rosettes. The colour range is restricted to browns and yellows, which I don't mind, but I shall never grow them again. Too many of the plants carried miserably small flowers and were dead-weight passengers.

At this point, in 1973, the Japanese anemones had been increased and rearranged in their informal groups and I planted in front of them, for that summer, the perennial *Anthemis tinctoria* 'Wargrave', treated as an annual. This is an extremely simple thing to do, as long as you've a stock plant somewhere from which to take your cuttings. The anthemis dies back in autumn to a low cushion of bright green mossy young shoots, which overwinter in that state. I took cuttings of these shoots in February. They can be packed 70 or more to a seed tray in a cutting compost and don't even require close conditions. The open bench of an unheated greenhouse will do. When rooted, in April, they're potted and so that they shall not become pot-bound before your spring bedding has been removed, they'll need a second potting into 5-in. paper pots and a strong compost like John Innes No. 3. They started flowering at the end of June and went on without a pause till late autumn. I dead-headed them three times. Each plant was given one stout cane for support and grew about 3 ft high. You can allow 2 ft between plants.

As they were too tall to come right to the front of the border I planted this area up with the annual bugloss, *Echium plantagineum* 'Blue Bedder'. This was sown in March, pricked out, then potted into 5-in. paper pots. They grew rather soft and lush, so I staked them individually before going on my Scottish holiday in June. By the time I returned they already looked like writhing green snakes and were rapidly running to seed. I had to make *ad hoc* arrangements for their replacement in early July, using the purple heliotrope 'Marine', from seed sown in early April, *Monarda* 'Beauty of Cobham' from plants I had lined out in a spare plot, and a few *Senecio viravira* to interweave with the others.

It is surprising how well most fibrous-rooted herbaceous plants will move in the middle of their lush summer growing season, if you soak them heavily before and after the operation. The monardas (a pretty, soft, pink-flowered strain with purple bracts) never flinched. I staked them individually after this move. Better safe than sorry. By September these ingredients had combined to make a pleasing display.

I must admit of the echium that it is an annual that gives a better account of itself on poor soil. It is especially good on chalk. Once established it will regenerate from self-sown seed, year after year. You can underplant stemmy roses with it. The blue strain, though harder to track down, is more effective than the mixed, which includes pink and white.

The next plot I hatched, though brilliant, was – well, I'm glad nobody's reputation except mine depended on it. I love yellow, especially in spring. I bought a lily-flowered tulip, 'Golden Duchess', and used as a carpeting background a foliage plant, *Barbarea vulgaris* 'Variegata'. *B. vulgaris* itself is a cruciferous weed of arable land. The variegated form has its glossy foliage handsomely splashed with yellow. It is one of the few plants having an unpatterned variegation that dependably comes variegated from seed. I sowed it betimes, on 11 April, pricked out the seedlings and lined them out so that by November, when I bedded them, they were very large. Too large, as it turned out. More than half the plants rotted in the winter. This was as surprising to me as it was disappointing. It is quite likely to be one of those subjects that becomes overlarge and lush if sown early. I should (most likely, but I'm not going to try again) have succeeded had I sown it a couple of months later.

For the summer of '74 the display was of *Penstemon* 'Drinkstone Red', raised from cuttings, with *Petunia* 'Blue Dandy' (an F_1 hybrid) in front. There were 110 penstemon plants from cuttings taken on 27 October. This sounds late, but even though they were potted off singly in early spring into $3\frac{1}{2}$-in. pots they were a little on the large side and rooting through when planted out on 22 May. This is a hardy penstemon with elegant narrow leaves and slender, not fat, funnels which you can grow as a border plant for several years, cutting it hard back each spring. But it'll never flower for half as long a season, in subsequent years, as it will in its first.

The petunia was a Multiflora type, which does not have the largest flowers but is comparatively weather-resistant. And I chose blue, first because you can nearly always rely on blue strains to waft the delicious petunia scent, of an evening, and second because, again, blue and pink petunias are far more weather-resistant than the reds. I sowed the seed on 7 April, pricked the seedlings out 7 by 4 (only 28 to a box), and they were perfect for planting out on 22 May, which is as early as you could want to bed out petunias almost anywhere in the kingdom. This does show how well one is served by late spring sowings.

The scheme remained viable until September when heavy rains put the petunias out of action.

For my next spring's carpet I used *Anthemis punctata* subsp. *cupaniana*, which is a vigorous, mat-forming perennial with greyish green leaves surmounted by a tremendous, seething mass of white daisies, at about 15 in., in May and early June. It is easy, with a fast-growing plant like this, to get busy too early, but I knew that and did not repeat my barbarea mistake. From one stock plant, the only one in my garden, I took every cutting I could find, and that was 195, on 10 August, potting them individually when rooted. They were planted out on 2 November and interplanted with an early-flowering blue Dutch iris called 'Ideal', 500 of them.

These irises were offered in two grades: 8/9 cm the smaller, 9/10 cm the larger. I chose the smaller – mistakenly. Very few of them were of flowering size, only about fifty, in fact, but the anthemis were dazzling and a feast in themselves, for the yellow eye in a white surround is a delightful contrast. I only had enough anthemis for three rows, so at the back I planted 21 plants of the perennial wallflower *Erysimum* 'Bowles's Mauve', which makes a large bush in a short time. It is raised from cuttings which I had taken, in this instance, the previous May, lining them out thereafter.

For the summer display I fell for a new, much fanfared strain of bedding dahlias called Redskin. The flowers are in a full colour range and are offset, in this case, by leaves of a dusky chocolatey hue. As dahlia leaves in general are distinctly uninteresting, this is a welcome method of pepping them up. Seed was sown on 12 April and the seedlings potted straight off into 3½-in. pots and J I P No. 1 on 9 May.

For the front of the bed I alternated large groups of *Tagetes* 'Cinnabar' with a curly strain of parsley, 'Bravour'. The parsley was treated similarly to the dahlia, being potted individually on 9 May. All were bedded out on 15 June; latish, because the anthemis flowered on into June, but in any case the late spring weather was desperately inimical until suddenly, on 6 June, we plunged into torrid summer and it stayed that way.

The dahlias' semi-double flowers ranged widely through yellow, biscuit and apricot to red, magenta, mauve and pink. As usual in seed-raised strains, some flowers were particularly good and worth saving as tubers for another year (I didn't), while others were particularly miserable and malformed and worthy of the bonfire. They started flowering in early July and made nice bushy plants. Of course they had to be dead-headed. It is so important, when doing this chore, to wear gloves. Otherwise you'll inevitably, sooner or later, be stung by a sleepy bee.

The tagetes, which is a bright copper marigold with single flowers and makes good-sized plants sufficiently uneven in height not to be monotonous, also benefits from dead-heading. You need merely to snap off the deads, which is quick and easy. With the dahlias you need to trace each stalk behind a dead head as far back as the next branch, for the looks of the completed job.

Parsley makes a brilliant green contrast to browns and oranges. Those of my plants that survived made splendid specimens, but half of them went down to a virus disease, spread by aphids, which first turns the foliage pink before killing the plant outright.

For 1976's spring bedding, I had sown a batch of perennial dianthus seed in the previous April, planning to treat these pinks as biennials, like sweet williams. The strain was called Ballet Mixture. They made beautiful grey-leaved plants, but I cannot recommend them as they included such a hotch-potch of pinks. Although described as an early-flowering strain, their June season was just when you expect pinks (in contrast to carnations) to flower. But some were so early and some so late that they didn't overlap. There were singles of the Highland hybrids type and these did not assort happily with a large proportion of doubles of a muddy mauvish pink.

Among them I planted my 'Ideal' irises, now in their second season, and this time they made a good display which started on 13 May and was over before the dianthus got going, but that didn't matter as the latter's foliage was a suitable background. Looking round my garden at the time the irises were out, I came to the conclusion that *Ranunculus gramineus*, which is at its best in May, would make an ideal carpet, if one raised plants of it in sufficient numbers. I have failed to germinate its seed but its tuberous roots are easily divided. This is an excellent dwarf, upright growing and compact buttercup with lanceolate, somewhat glaucous foliage and very clear but not harsh yellow flowers.

1976 was the drought summer, it will be remembered, so the pinks' season was curtailed. I replaced them on 10 July with a strain of mixed bedding dahlias sown on 1 May and subsequently kept growing strongly by potting into 5-in. black paper pots. These bedding dahlias never let you down, but really their mixtures are too mixed and include (in the same way as antirrhinums) too many warring colours. Another time I shall buy two or three colours in selected strains (only obtainable from wholesale seed merchants, nowadays) and mix them when planting out. I might even grow 'Coltness White' dahlias. Would that look too self-conscious and affected in front of white anemones? I don't think so. (Since I wrote that, the single colour strains of dwarf Coltness Series dahlias have disappeared from even the wholesale lists, all except the scarlet and the yellow, which are too hot for me.)

I wanted next an underplanting of yellow pansies in spring 1977, with a super Dutch iris that had only recently come down to a reasonable price level, 'Purple Sensation'. I sowed the pansy, 'Golden Dream', on 2 August. Pansy seed germination can be tricky in hot weather and we were still at the height of one of the hottest summers on record, so I placed the seed box, covered only with a pane of glass, outside, under some ash trees, and watered it frequently. Germination was excellent. The seedlings were pricked out 7 by 4 and 228 of them bedded in October.

However, we all know what happens to the best laid plans. The 300 iris bulbs got lost en route. By the time the second consignment arrived we were half-way through November, some 16 inches of rain had fallen in the previous ten weeks, and the ground was a pudding when we somehow or other got the bulbs in. Most of them rotted. Those that flowered revealed that they were not 'Purple Sensation' at all but some sort of mixture. The pansies did well but looked a little foolish on their own.

For the summer of 1977 I decided to revert to an old favourite that I've used only once in the last thirty years but which we frequently grew here when I was a child: *Phlox drummondii*. The large-flowered strains of this annual have a most inconvenient habit. They make such long stems that you either have to support them in an upright position with twigs or you have to pin them horizontally to the ground with hairpins or bent wire. So I grew the smaller-flowered Cecily Group which makes a bushy plant.

Sown in early April, it started flowering in July and really got into its swing in August. It faltered during bad weather in early October but was radiant again for the rest of the month when beautiful weather returned and I was quite loath to turn the plants out on 17 November. They had served me well but the criticism of too many colours in the mixture applied. In the mass they blurred.

Sweet Williams next, interplanted with replacements of the 'Purple Sensation' iris. I grew two sweet William strains, a very dark red ('Morello') and a white, and planted them two dark to one light. I was planting three rows concurrently, and as all the plants looked more or less similar (I drew them from two separate barrows) I found that the easiest way of obtaining an even colour distribution without forgetting which I'd planted where, was to place a sprig of white hydrangea blossom on top of each white plant as it went in.

As usual, even when you buy the larger grade of bulb (and this is essential to get any bloom at all) only an unsatisfactory fraction – perhaps half – of the irises flowered. It is probably best to grow your bulbs every year in nursery rows. When harvesting and sorting at the end of the season, use only the largest for bedding and line out all the remainder for the following year. Flowering weakens a bulb so that it takes a year off, but those that have had a year for recuperation plump up again for the following year. None of this is noticed if you grow these irises in permanent plantings here and there in the garden. Then they'll always seem to flower freely and those that are resting will not be noticed among those that are performing. There may be a moral here but I shall continue to ignore it.

The red and white sweet Williams were smashing. At their best when I was in Scotland in July, alas, but never mind; I saw them before and after and the proportions 2:1 of red to white were just right. Although 'Morello' is a dark red it's not so dark as to be ineffective at a distance, as 'Nigrescens' would be.

To follow these I had got the Dutch students who were working for me to sow nasturtiums singly in 3½-in. plastic pots, using JIP No. 1, in mid-June, and these were stood outside beneath an irrigation spray line. Seeing that the Japanese anemones have white flowers with yellow stamens, I fancied yellow nasturtiums. I liked the idea of them filtering into and climbing among the anemones but I didn't want them to stretch forwards on to the lawn. So I used double 'Golden Gleam' nasturtiums behind and 'Whirlybird Gold' in front. The latter is a bushy strain without any trailing aspirations. Its flowers have no spurs. Germination was rather poor in its case, and the planting looked a bit thin at first but not for long.

The changeover was made on 1 August between fantastically heavy thunderstorms, and within a week the 'Whirlybirds' were in flower. Ideally I should have sown them no more than one calendar month before planting out. As it was the plants had grown into each other and flopped about in a slapdash manner when planted. But all soon righted themselves and knit together.

The 'Whirlybirds' were so free-flowering and so comparatively short of foliage that I thought they would soon run out of steam. In fact they kept up the pressure till the very last. The 'Gleams' have the larger, handsomer, more intensely coloured and substantial flowers, but fewer of them and with the common nasturtium habit of making too much leaf, but that was only in the earlier stages. There were two red rogues which I pulled out, despite the gaps they left. This, of course, was only momentarily noticeable.

After a miserable summer, the weather suddenly settled down on 17 August and it was the finest, driest autumn I can remember, even right up till the last week in November. We finally hoicked the nasturtiums out in the second week of that month. They'd done us proud for fourteen weeks.

I had planned for antirrhinums in 1979, and sowed these on 9 August; subsequently pricked them out and then potted them into 4-in. long tom plastic pots for overwintering in a cold frame. They were strong plants by the time the cold weather finally arrived and I was prepared to dispense with spring bedding so as to be able to plant the snapdragons out in April.

However, I relented. Originally intending to grow 500 of the very early flowering *Tulipa kaufmanniana*, which should, in a normal year, have finished flowering by early April, I was so late in ordering that I had to take what Peter Nyssen had left, which was some 450 of the Kaufmanniana hybrid, 'Shakespeare'. They were planted under ideal conditions in mid-November.

They were just right for Easter on 15 April but early tulips are a luxury. Bad weather at flowering time destroys them sooner rather than later.

My antirrhinums were of the F$_1$ Sprite Series strain which they still (bless them) sell in separate colours so that you don't have to endure pinks warring with yellows, etc. I chose four colour strains: 'Orange', 'Scarlet', 'Yellow', 'White'. They are dwarf without being too dwarf and carry handsome spikes at some 18 in. or so. They had overwintered well and were very strong plants. This way of raising antirrhinums without heat at any stage (and it had been the coldest winter since 1963) is extremely satisfactory. However, we were not out of the wood. We planted out on 3 May and already, then, there were spots of rust showing on some leaves. The writing was on the wall. We soaked them at planting out, with a benomyl drench, but to no avail. They succeeded in carrying their first flush of blossom in June but by July had to be scrapped.

How it is that the rust always reaches me so early, I wish I knew. Some gardens escape it until so late in the season that it doesn't matter, or even altogether. We hadn't grown antirrhinums for years and had no volunteer perennial plants hanging around and our neighbours are a long way off, so???

Anyway, seeing that we'd had early warning, we pricked out masses of seedlings of a newish mixed heliotrope strain, 'Regale', and subsequently potted them individually in 5-in. paper pots, so they held well until required, from a sowing in the second week of April. It is a useful attribute in heliotrope when you need them late, that they are quite slow to develop in the early stages. They were planted out on 16 July.

Their display continued into November. Ideally I would not make a large planting of cherry pie on its own, as the mauve and purple colouring is apt to be sombre on a grey day. The scent was not nearly as strong as in some of the old cultivars that have to be maintained from cuttings and overwintered in a well heated greenhouse, but nice whiffs came off 'Regale' from time to time.

I had planned to follow them with wallflowers but we were plagued with rabbits that summer and lost all our wallflower and lupin seedlings.

I had strong plants lined out, originally from a sowing in April the year before, of *Geum borisii*, and we also had large clumps in a nursery row of the purple *Viola cornuta*, so we split both these and bedded them out in alternating patches which were interplanted with 298 (my dachshund ate two bulbs) of the late-flowering orange tulip 'Dillenburg', an old favourite. Purple and orange.

For reasons which I cannot explain with certainty, neither the viola nor the geum flowered at all freely in the spring of 1980. The splitting and mid-November planting may have disagreed with them. Even likelier, the trouble may have been that, after planting, we treated the area with Venzar, which is a (now illegal)pre-emergence weedkiller based on lenacil. This often has a stunting or retarding effect on perennials that have not become established previous to its use.

My next plan also went awry. I intended, as earlier intimated, to have white 'Coltness' dahlias, but my seed in its second year (fresh seed being now unobtainable) was no longer viable so I made groups of various plants I had available.

I need not go into details. The results were pleasing if a bit of a hotch-potch. I have now covered a long enough period to show, I hope, that, first, my plans are as likely to run into trouble as anyone else's but I'm not unduly worried about that. Second, and most important, there's a tremendous kick to be enjoyed from this game and when things do work out well I feel as pleased as a dog with two tails.

THE GARDEN IN THE HEAT

No gardener reads books on his subject during a heatwave. Either he's trudging around with watering cans, moving spray lines and sprinklers, or else relaxing with cool drinks and conversation. The time to be reminded about those hot days is in shivering January when I have time to write (snow without as I do so), you to read and a bit of vicarious pleasure may be in order.

There are certain times of the year when a pattern of weather tends to become established which repeats itself over and over again for several months. If an endless chain of depressions, with the wet and the wind and overall mouldiness that attend them, is operating in early July, you can more often than not say goodbye to summer as we understand the word. But if, at the same season, a fine spell has no sooner broken than it is (catching the met. men on the hop) succeeded by another fine spell and that by a third, then we're probably in for one of those summers that everyone longs for and even expects (forgetting that we're not in the Mediterranean) but that rarely actually occur in England. Not all that rarely, perhaps. We, in south-east England, can look back on 1947, 1949, 1955, 1959, 1964, 1970, 1973, 1975 and 1976, which works out at one good'n in three over a thirty-year period.

Two warm days in succession will produce a stream of complaints from nearly everyone I meet, and yet, basically, most of them believe that summer should be summery and that there is reason for satisfaction when it works out that way. It is not the midday actuality of heat that we enjoy (and gardens look pitifully flat and flabby under the glare of a high sun) but its legacy: the warmth combined with freshness of evening, morning, night; the scents.

Gusts from *Genista aetnensis*, the Mount Etna broom, throughout July; and the late Dutch honeysuckle, *Lonicera periclymenum* 'Serotina', although scarcely visible within an enfolding cherry, flings its scent half-way across the garden from night till morning. *L. japonica* 'Halliana' is even stronger, though more cloying; sticky-sweet. Above all there is the summer jasmine's incomparable fragrance. Overpowering, yes, and yet I can never have too much of it.

The pungent sweetness of border phloxes is essentially a daytime pleasure. It seems that one can never speak or write of phloxes without launching into the problems they engender or epitomize. In hot weather, above all, it is the water problem. You can judge a garden by the state of its phloxes. They are a pitiable sight when parched, a living reproach to their owners, to the country's water resources and, in the background, to an entire social structure which fails to provide an environment fit for phloxes, and to an educational system which fails to din into obstinate blockheads that an adequate water supply is essential to good gardening.

The phlox (as also the hydrangea) is symbolic of all this, wearing its heart on its sleeve for all to see. But consider, now, the reverse of the medal, the happy phlox. Those glorious soft, fat panicles of moon-shaped blossoms, breathing their message of summer's opulence. Tall phloxes and short; purple, pink, white, mauve and magenta phloxes in great pools of colour and of scent. With phloxes for contrast, you can grow any number of foliage plants in your mixed borders, the one setting off the other. And all that these phloxes need is water and manure, the good things of garden life.

It is ironical that the tally of plants we can most enjoy in hot weather includes so many moisture lovers. Moisture makes for luxuriance and luxuriance in a plant conveys a sense of well-being to the onlooker.

But there are other plants, perhaps more appealing to easygoing gardeners, that simply want a baking to give of their best, either now or later. *Zephyranthes candida*, the flower of the west wind, responds to a hot summer immediately. Its 6-in.-tall white crocus-flowers begin to appear in early August and make a tremendous show through the next two months, given this inducement.

The hardy agapanthus should flower freely in most years. With their fleshy white roots they are able drought resisters. August is their main season, but although they have become popular and are easily grown it is surprising how seldom they are really effectively used. Blue is a colour that needs offsetting, but with what?

The agapanthus globe head contrasts well with spikes and I have taken pleasure in the deep blue 'Isis' (but remember the deeper the blue colouring the smaller the flowers) with the poker *Kniphofia* 'Modesta' which has cream and coral spikes on a 2-ft plant. In front of them some plants of the non-stop-flowering, acid yellow-green *Euphorbia seguieriana* subsp. *niciciana*. Agapanthus should look well with the red forms of *Crocosmia masoniorum*.

That sumptuous climber *Campsis grandiflora* (*C. chinensis*), which arouses our envy on continental visits, usually succeeds in opening and developing its flower buds only once in ten years in southern England (never, any further north). But in really red-hot years like '75 and '76 it is already flaunting before the end of July and will have the chance of unfolding its entire quiverful of buds over the next couple of months. They are carried in large terminal panicles and open into plump orange-red tubas (rather than trumpets).

Flowers with southern blood in their veins that often fail with us altogether, do, given that outstanding summer, get cracking phenomenally early, just as though they always did it and what's all the fuss about and who was murmuring under their breath about throwing them out for not paying their rent? The clematis 'Lady Betty Balfour' (and she's not the only one), purple-flowered with white stamens, just makes a lot of soft green unproductive growth in chilly years. Given a reasonable season she's a September flowerer but after a stinking hot summer like '76 she's blooming by early August.

All of which makes you wonder what there'll be to follow, and yet you never in fact find yourself with a blank on the arrival of autumn. On the contrary, many plants that would normally give you only one crop will have stored up so much energy as to be able to produce a second. Herbaceous phloxes are a typical case in point, likewise *Campanula lactiflora*, *Salvia* × *superba* and the earlier-flowering heleniums.

Roses are reduced to pulp by scorching weather in June–July, but if they have it in them to be repeat-flowerers at all, just wait for the August–September display and then go on admiring it into October. Like many people, roses enjoy heat in retrospect more than at the time.

In writing of a hot summer one must mention its legacy beyond the season. The pure yellow 'crocuses' of *Sternbergia lutea* will start flowering and abundantly at that (if you start with a free-flowering strain) in mid-September, whereas in chilly years their feeble attempts will centre on November and even up to Christmas. *Amaryllis belladonna* (the very name makes this captious bulb worth attempting) will throw a forest of its thrilling purple stems, soon to be crowned by clusters of scented (rather a cheap scent, I'm bound to say) pink trumpets in September–October. Later we shall have quantities of *stylosa* irises (*I. unguicularis*) to pick through the winter, and most of your spring-flowering shrubs should be encrusted with buds to keep you (and the bullfinches – but at least they leave the magnolias alone) in a state of pleasant anticipation throughout the winter.

One mustn't forget, even in the depths of winter (my bay tree is covered with a film of ice from rain that has frozen on it and when a branch sways it lets out an eerie creaking noise) that hot weather is not all beer and skittles. OK if you can just sit or swim around, but if there's work to be done it will go forward very noticeably more slowly in the heat than when it is cool and bracing (although I loathe being braced). Feet ache, tempers are short, the potting shed is an oven and is shunned, although cuttings are queuing to be taken. By six o'clock one can breathe again and the world becomes transformed.

In a settled summer there is a feeling of continuity. You find yourself drifting through a long succession of warm days, sometimes sunny, sometimes (for a welcome change) overcast, but dependably repetitive. The season develops a drowsy rhythm and you expect it to go on indefinitely. 'Here,' you feel, 'is the reality of summer in England and I am there, at its centre.' That is good.

PEOPLE, PLANS
AND PLANTS

PLANNING A BORDER

Making a border plan is a great excitement; most of all, of course, if it's for yourself, but exciting also for whomever it may be. Because, for one thing, the opportunity is rare. Most of our garden alterations and additions are piecemeal over the years, a gradual implementation of new ideas, experiences and plants. The changes are almost imperceptible and, retrospectively, it is difficult to remember how the scene appeared before the alterations were made. It is well worth making photographic records; indeed it is encouraging to do so as you then have a yardstick wherewith to measure progress – in the plants' growth and in your handling of the plant material.

This slow and steady evolution is fine; something to live with and a part of your life. But it is of quite another order from the plan; the sudden creation of a completely new feature where there was nothing, or something very different, before. I know a border doesn't spring into existence just by the wave of a wand (though many impatient garden owners think it should happen that way). It takes time from their planting for the ingredients to make their mark – only weeks if they are annuals, a couple of summers for hardy perennials, three to five years for the majority of shrubs. Even the shrub's time scale is not all that long, and the whole evolutionary process is an immense thrill. The original plan was like (why be modest?) the creation of a universe by a sudden explosion where before there was only gas.

The gas or chaos in our case takes on the familiar horticultural guise of bindweed, ground elder and couch. There are material advantages in clearing a border site completely before making a new start. And in making sure that the site really is clear. Take an extra year about it, if necessary, with a varied range of weedkillers designed to catch every variety and type of weed on the hop one way or another.

But however far you've progressed in your border preparations, the time for planning is in winter while the snow is on the ground. First you make lists of the plants you want to include. By the time you've eliminated a great many of those your eye lights on in seed and plant catalogues, in nurseries and garden centres, your list will be, perhaps, a hundred strong. And it's only a reservoir of possibles to draw on. When the plan is actually drawn up it will, even for a large border, be reduced to fifty or so.

Because this is one of the rare opportunities in a plantsman's life for *not* overcrowding. This once, given a bare expanse, you can be firm with yourself and insist on including a high proportion of large, bold groups of single varieties in contrast to your preference for popping in one of this and one of that. There should be singletons, most certainly. A single fountain of a grass like *Miscanthus* or a specimen bamboo or an imposing shrub like *Olearia macrodonta* will be more impressive where numbers would confuse and weaken, but where colour is more important than form, as it often will be, then you'll want large patches.

To consider in a little more detail the factors that will decide what goes into your list of possibles and what's left out. Clarify your mind first of all on what sort of border you wish to aim at. You must restrict your aims, otherwise it'll fall between any number of stools and be effective only in its incidents, never in its entirety. So the main thing, obviously, is not to try and make it cover too long a season of interest. The smaller the border, the shorter the season it should be required to grace.

I know you can include shrubs – evergreens, most likely – that will look good the year round and for the same reason you'll include plants whose foliage is a positive asset for months rather than weeks, but that will still leave your show groups, the phloxes and poppies and salvias and roses of this world. You do want to have a definite idea of what your border's high season is to be. Diverge from that if you like, but so that your divergences will in no way detract from the main idea. For example, in a border concentrating on a late summer season you can still include tulips, but in such a way (as by planting them among later-flowering perennials) that they will not leave a blank when their season is past.

You'll find from your own and other people's experience that the season for which you can plan the longest powerful display is late June till (if you're resourceful and the border is a large one) September.

If you decide on a shrub border or a herbaceous border, that'll be a further limitation on your choice of materials but, in my opinion, an undesirable one. Shrubs and herbaceous plants go well together, the one most often contributing substance, the other colour, and both making rather different contributions in form. Take advantage of their complementary assets and make it a mixed border, adding bulbs, annuals, tender perennials and anything else that takes your fancy if you consider that they will make a positive contribution to your overall plan.

Soil will be a limiting factor. You'll probably know what your soil will or won't grow easily before you start, simply by looking around you, and you'll make allowances according to whether it's heavy or light, acid or alkaline. Climate and aspect. So long as the moisture is there, a shady border need cause no dismay, for there are a great many plants that revel in damp shade, but if trees and tree roots are mopping up all the food and drink as well as light you'll be severely handicapped and should not be too ambitious. Sunny, protected sites invite a bit of risk-taking with plants that need to be sun-soaked if they are to flower freely. Here, too, your grey and glaucous foliage plants will develop their best colouring.

Mottoes and themes. Some gardeners like to simplify the bewildering question of what to plant by deliberately imposing rules and restrictions on themselves. They don't think of it like that, but this is nevertheless very often the reason for devoting borders to all of one type of plant or to all of one or two colours.

Then there arises the question of availability of material and cost. Don't include plants just because you have them or have been given them. That doesn't *per se* mean they're suitable. Make them run the hoop, by yourself being their severest critic in respect of their possible role in the kind of border you have in mind. For any given position or juxtaposition there will be many suitable alternatives. The excitement is in deciding which one, but you needn't be ashamed of frequently changing your mind.

The fact that you may need to fallow your border for an extra year before planting it because of a legacy of weeds can be turned to advantage so long as you have a spare plot in which to grow plants on. Don't consider this a heeling-in exercise so much as a lining out. Properly planted and looked after by watering them in dry spells, many herbaceous plants will so increase in one year that you'll be able to pull them to pieces at the end of it and multiply the numbers available for border groups by five or more. That'll save noticeably on bills. Likewise, by growing your shrubs on, you'll be able, when the time comes, to plant up with units that you can see and that won't look utterly lost. None will mind a second move at the end of a year; they'll be used to it. Allow plenty of space, even though it'll look exaggerated to begin with, between a shrub and its next neighbour. Six feet between shrubs is fairly average; 3 or 4 ft between a shrub and the first herbaceous plant in a neighbouring group. Even then you'll have to guard against lush herbs crowding a small young shrub during the growing season.

I like to make a plan on squared paper to a large scale: 1 in. to 3 ft is ideal for allowing you to mark in individual plants at a suitable spacing within a group so that you have a shrewd idea as to how many will be needed. If you've a good stock of your own already, this won't matter; you can then decide how many plants or pieces to use when you're laying them on the ground ready for planting, but if all's to be bought in, numbers are essential.

When you sit down to make your plan the blank paper in front of you and the long list of plants by your side look daunting. But once you've outlined something important that you fancy on a salient bulge or corner, you'll quickly see in your mind's eye (or remember from having seen the same effect elsewhere) a plant or planting that will make a handsome neighbour to the first one. And then a companion for those two and so on, bearing in mind as you cover the plan any repetitions that you may have previously decided will lend purpose and cohesiveness to the scheme. Not many of these are necessary. At least I think not, speaking as a plantsman who enjoys variety more than repetition, but a few repeats, given a border of any size, yes.

At the end of it you'll find there are certain plants you meant to include and haven't. I shouldn't let that worry you. The great thing, as I said at the start, is to make this initial plan one in which at least half the groupings are large and bold. There'll be all too much whittling away of group size in the years to come when (if you're mad about plants) you'll feel impelled to find room for treasured newcomers. If, on the other hand, you're an intellectual gardener for whom the importance of plants is in the role they perform in a larger scheme rather than what they look like and how they behave as individuals, this problem won't worry you. I shall visit your garden, but only once!

Transferring your plan on to the actual site is easy, though less simple, of course, if the outline is irregular. You need a yard-square (or metre-square) grid on the ground that corresponds with the grid on your plan. Equip yourself, therefore, with lots of straight sticks, 18 in. or 2 ft long, that you can push into the ground to mark the corners of each square. Pieces of bamboo cane are ideal, but maiden growth from apple, pear or plum prunings or the year-old stems from coppiced willow or hazel will serve admirably.

Then, using a nice long reel tape, you start along the front of the border if it has a straight edge or from any known datum point if it hasn't, pushing your sticks in at imperial or metric intervals. Then take a line at right angles to the first where the border is widest (if its width varies) and mark that out. Then back to your original plane but as far from the first line as you can get (this will be the back of a regular oblong border); mark this line out and gradually fill in your grid. The grid *can* all be done with string, if you've plenty of string, in which case sticks will be needed only at the perimeter to keep the string in place. But I think the string is a bit of a nuisance when you come to the actual planting.

Your group outlines are marked in the plan and these you now transfer to the ground using the sharp end of a stoutish stick (a walking-stick isn't quite stout enough). With the help of your grid, this can be done quite accurately. Now you want to emphasize the outlines by dropping some material that shows up well into the channels. Don't make Hansel and Gretel's mistake of using crumbs of bread! The birds will gobble them up. I find that horticultural grit shows up well on my heavy soil, but if yours is sandy you may find that peat stands out better.

All this marking out should be done at the last moment before planting so that your outlines don't have time to be blurred by the weather. Lay out all your plants in each group before interring them and in doing so try not to be so clumsy with your feet, fork and trowel as to obliterate group outlines you've not yet dealt with.

PATHS, PLANTS AND PEOPLE

A friend kept telling me that I ought to write about paths, Eventually I asked her if she had any special aspect of the subject in mind and she mentioned the importance of making garden paths take the right route, so I will start from there and see where it takes me.

Which way shall a path run? If it is a much-used service path leading, say, to a house door or to the compost or rubbish heap it is essential that its route should be direct. I mean, if there's a corner that can be cut you may be sure that it will be, if not by you then by everyone else on the property, so it's best to take the line of least resistance and make the short way official.

If this would spoil a formal layout and you feel that a right angle is obligatory, then you must make physical barriers that will prevent corner-cutting. Remember that dogs, having no aesthetic and few moral standards, will cut corners that humans wouldn't dream of desecrating. Prickly shrubs like berberis are effective. In my part of Kent and East Sussex you can still buy miniature hurdles standing 2 ft high, and I find these most helpful for protecting a corner even if only temporarily while, say, there is snow on the ground making corners invisible or while young plants are growing up that will presently be of a size and bulk to look after themselves.

Paths that curve informally on an uneven piece of ground, edging their way round a tree trunk here or a bush there, can look charming and inevitably right. Either their route (especially in the case of a foot-trodden-and-created grass path) has made itself from necessity and common sense, or the constructor has had an eye and proper feeling for how these things should go. We have a narrow brick path that takes a wide curve through our orchard. It breaks the expanse of turf with a meandering and yet purposeful air. Aesthetically it is satisfying and yet this is very much of a service path. One would, and usually does, prefer to walk on the kinder grass itself, but when all is soggy and a heavy barrow has to be propelled, the bricks come into their own. In late summer and autumn, the curve in this path does not seem to argue conclusively against short cuts, but when the daffodils are up and the grass subsequently becomes long, the path is a must. At all costs, though, avoid meaningless kinks and wiggles that are obviously intended to suggest an informality that is actually contrived and spurious.

Much-used paths must be more durable than turf. If they are set in turf – say on the stepping-stone principle – or run alongside turf, make sure that the level of the path is slightly lower than the turf. Then you can run the mower around without anxiety and the turf can abut right on to the paving. After months rather than weeks, the grass will tend to grow horizontally over the paving (especially if the turf includes creeping grasses like *Agrostis canina*) and you may have to trim with a knife along the stone edge, but this is child's play compared with the other alternative. When paving is higher than turf, the latter is cut back from the former with a half-moon edging tool (an implement of destruction that should be banned). Soon an ugly gully develops between lawn and path and there is an unceasing problem of how to keep the gully weed-clean.

What materials you should choose for a paved area or path is a very disputable question, and one largely governed by what you can afford. Unless the area to be paved is very small, square blocks look unbearably monotonous. Try and use rectangular blocks of varying sizes, but not of varying colours. That's too exciting, which means distracting and that you're placing a greater emphasis on your paths as features than they merit. They are, after all, subservient to other garden features. When you lay your blocks, aim always at T-junctions in your cracks. Never allow cross-roads. The former are more natural.

Concrete paving, well made, can be very acceptable nowadays, and it is less slippery than York flagstones. But there is concrete and concrete. One variety allows water to soak through it so that you get no puddles. It looks peculiarly lifeless, in my eyes, and I've only seen it made in squares. Brick paths made of old bricks set on edge (like that going through our orchard) can look most attractive, but they are prone to slipperiness, especially in shady places. Crazy paving, with its uneven, ye olde character, I nearly always detest, but that's probably just me. Gravel is cheap and can be edged with bricks at a slightly higher level so as to prevent it spewing on to beds and lawns. Ordinary yellow gravel, composed largely of rounded stones, moves around and needs frequent raking. Sharp-edged, grey stone gravel, such as I meet in Scotland, is really beautiful, however.

What I enjoy more than anything is to see a piece of homemade, semi-formal paving done by an amateur with a feeling for patterns and using a variety of materials. I wish I could tackle that sort of job. There is a fascinating example at Oteley, a garden in north Shropshire. Patterns are made with flat grey stones set on edge, forming the background, and rounded white stones, describing rectangles, straight lines and arabesques. They still have the wooden S-shaped template that was used for describing the figures of 8, when the paving was laid down about 1855. The grey stones came from the bed of the River Dee and the white is quartz, found in the garden's own soil.

An important question arises when a satisfactory width for your paths has to be determined. Let us suppose that you have a piece of naked ground in which you want to make a border (or two borders). Whether formal or informal in outline, make it as large as you dare. Otherwise, in later years, you will regret the lack of elbow-room. (In parentheses I should give credit to another point of view which Margery Fish shared: that no border should be wider than allows you to reach and weed its contents without stepping on it. This is the cottage garden principle. I prefer to think big even on a small scale.) If the border is going to look comfortable in its surroundings the plants around its margins must be able to spill over the edges. To realize this at the start will save much heartache later. It means that the border's effective size will be greater than it looks on paper and this in turn means that if the border is adjacent to a path, the path itself will need to be wider by 2 ft or so than you might have expected had there been no overspill.

In our family a bachelor's path is one that is only wide enough to allow people to progress along it in single file. This is all very well if it is merely a question of wheeling a barrow from A to B; not so, if you want to saunter in a companionable fashion. I will give one example from a garden that I know where the proportions strike me as being exactly right. The path between a pair of rectangular borders leads to and is ended by an important gateway. The borders are 12 ft deep, 30 yd. long. The path between them is paved and it is 7 ft wide. Seven feet may seem an awful lot, but it is not, in fact, an inch too much by the time mat formers like *Cotoneaster microphyllus* have taken their share.

If you cannot afford to pave a large area like this, but have to grass it over, do at least pave a 2-ft margin between the border and the grass.

Supposing you have a one-sided border backed by a hedge or fence but fronted by a lawn, it is an admirable plan to have a 4-ft wide path that will allow the passage of two abreast and also act as a divide between border and lawn. If you grow compact plants in the front of your border, 4 ft will be wide enough, but if, as would always be my inclination, you allow pool formers to flow forwards on to the path, then you had better allow a 5-ft width of paving. Even then, you and your companion may sometimes, one or other of you, find yourselves nudged on to the grass and your progress will take on a drunken imprecision. Worth it, I think.

In the best-regulated gardens, weeding between paving stones is never necessary. Either the cracks are cemented in, or the application of modern weedkillers at the start of each new growing season does all the work for you. That is the labour-saving approach to the problem, but I have no wish to visit the gardens in which it is relentlessly practised. They are as soulless as they are efficient. Weedkillers are invaluable on paths that are frequently trodden, but wherever you have an extensive area of paving, for instance on a terrace where you sit out or surrounding a formal pool, the happiest treatment will be to introduce natural crack-lovers of pleasing appearance (and some of them will introduce themselves); then to allow them to get on with the job of seeding themselves (for example, blue-eyed grass, *Sisyrinchium angustifolium*) or of rooting as they trail (such as *Acaena novae-zelandiae*, which carries crimson burrs in summer).

The corollary to this humane and relaxed form of gardening is, of course, that there will be weeds seeding in as well as the plants you want to encourage. They will have to be dealt with and, nine times out of ten, the best way of achieving this will be by getting down on to a comfortable kneeling mat after a deluge and, with a sharp-edged pointed trowel or a pruning knife, to get slicing and tugging. It is surprising how quickly the work goes. When a tough-rooted weed – perhaps a year-old seedling ash, sycamore or bramble – has to be extracted, don't go at it with a sharp, straight pull; it will break off and come again more strongly. Gently pull, and at the same time lever to and fro along the crack. It is wonderfully satisfying to get out bramble seedlings in this way, and you will perennially find them and elder saplings coming in wherever a tree of any species overhangs your paving. Black-birds are the culprits; I sometimes wonder why they should be my favourites.

ANOTHER LOOK AT MISS JEKYLL

It was surely time that Gertrude Jekyll's two masterpieces, *Colour in the Flower Garden* and *Wood and Garden* were reprinted. Now at last we have them. There is so much to learn from them, and a lapse of seventy or eighty years has not staled the lesson. For Miss Jekyll knew how to manage plants, culturally, and also how to manipulate them for effect. She teaches us that gardening of the best kind is not easy, and that there are no short cuts, but she also fires us with her own enthusiasm. We can clearly see, after reading her, that the effort is worth while.

An inspiring facet of this great gardener is her lack of stereotype. Her name is associated with the herbaceous border and this garden feature is generally acknowledged to be outmoded, requiring far too much labour. Actually, the island beds promoted by Mr Alan Bloom are herbaceous borders under another name and using modern cultivars of moderate stature. But the point I want to make is that Gertrude Jekyll didn't care twopence whether the plants in her borders were herbaceous or not as long as they contributed what she wanted of them. She did not refer to her herbaceous borders, ever, but to her hardy flower borders or mixed borders, and then generally in inverted commas because they did not fit in with the general conception of such features. 'I have a rather large "mixed border of hardy flowers",' she wrote. 'It is not quite so hoplessly mixed as one generally sees, and the flowers are not all hardy.'

On the point of not over-mixing, she shows the artist in her. She was an artist, first, by training and until her eyesight began to fail. 'Planting ground is painting a landscape with living things, and I hold that good gardening takes rank within the bounds of the fine arts, so I hold that to plant well needs an artist of no mean capacity.' Of a large border, then, she says it is important 'to keep the flowers in rather large masses of colour. No one who has ever done it, or seen it done, will go back to the old haphazard sprinkle of colouring without any thought of arrangement, such as is usually seen in a mixed border.' 'The next flagrant fault, whether in composition or in colour, is the attempt to crowd too much into the picture; the simpler effect obtained by means of temperate and wise restraint is always the more telling.' And again, 'There is nothing much more difficult to do in outdoor gardening than to plant a mixed border well, and to keep it in beauty throughout the summer. Every year … I find myself tending towards broader and simpler effects, both of grouping and of colour.'

I don't think she loved her plants as individuals any – or, at least, very little – less than the plantsmen among us who give their all to cultural details but have no idea of presenting their material.

Miss Jekyll's main border was 200 ft long by 14 ft wide. It had full stops at the ends and two-thirds the way along (where divided by a cross-path) made with big yuccas, bergenias (*Megasea*, in her day) and *Stachys byzantina* (syn. *S. lanata*). The border had a colour scheme (not slavishly adhered to): cool at the ends, building up to hot in the middle. Thus at the west end it starts with blues and greys merging into pale yellows, then brighter yellows, orange and red. From there it cools off again in the same order except that at the east end there were mauves, pinks and purples rather than pure blues.

On the whole she preferred colour harmonies to colour contrasts. I think most of us come to this preference as we get older. But her exception to this rule was blue. 'Pure blues always seem to demand peculiar and very careful treatment.' Thus she liked a full blue, as she called it, with a pale yellow but couldn't bear blue with mauve or purple. For this reason she had the mauves and purples at one end of the border and the blues at the other, that is delphiniums, anchusas, *Salvia patens*, blue Cape daisy (*Felicia amelloides*) and lobelias. She also tolerates *Campanula lactiflora* and *Clematis tubulosa* (syn. *C. davidiana*) here, but only because they flower when the delphiniums and anchusas are already over.

Nearer the centre of the border we find *Eryngium × oliverianum*, which is pure blue, set in the middle of contrasting yellows: primrose African marigolds, tall yellow snapdragons, double meadowsweet, *Achillea filipendulina* and coreopsis. Her plan, included in *Colour in the Flower Garden*, is fascinating to study, although clearly it didn't stay the same from year to year.

She reckons that the height of her border's main season does not start till the second week in August. Indeed, of early July, taking the garden as a whole, she writes: 'After the wealth of bloom of June, there appear to be but few flowers in the garden; there seems to be a comparative emptiness between the earlier flowers and those of autumn.' Few of us would agree, judging by the number of gardens that choose to open to the public at this very moment. Speaking for my own, I reckon that it peaks in early July and holds it for the next month, falling off steadily from the second week in August.

The reason for the difference is not hard to pinpoint. Miss Jekyll chose to have it that way. She liked a delayed climax and she achieved it by a wide use of annuals, short-lived perennials (e.g. penstemons), *tender* perennials (dahlias, cannas, *Salvia patens*) and yellow-flowered composites: helianthus, helenium, rudbeckia and coreopsis. She also plunged a large number of pot plants, where a part of the border had turned dull. Indeed, if she wanted a pink hydrangea, say, in a position where something else was already growing, and provided the latter was an easy-going and tolerant perennial, she would have a piece of it chopped away to make room for the newcomer. Besides hydrangeas, *Lilium longiflorum* and *L. auratum*, *Campanula pyramidalis* (always unsatisfactory as a garden flower, in my opinion) and *Plumbago auriculata* (syn. *P. capensis*) were introduced.

'I have no dogmatic views,' she tells us, 'in having in the so-called hardy flower border none but hardy flowers. All flowers are welcome that are right in colour and that make a brave show where a brave show is wanted.'

How sensible! It is only the timid gardener whose mind-forged manacles constrain him to think entirely in terms of categories and compartments for different kinds of flowers in different areas of the garden.

To get the red she needs in the centre of the border Miss Jekyll uses cannas, hollyhocks, dahlias, scarlet salvia, red celosia (the cockscomb), scarlet and orange nasturtiums, penstemons, gladioli. The only really permanent ingredients are red hot pokers which she calls tritomas (kniphofia), *Lychnis chalcedonica*, *Lilium lancifolium* (syn. *L. tigrinum*), the double form of *Hemerocallis fulva* (which is tawny, not red) and a border phlox. Some of these only flower for two or three weeks whereas the dahlias, cannas, salvias, celosias, penstemons and nasturtiums would contribute for many weeks and at a late season.

It is therefore worth pausing to remark that those who would plant a red border or red garden should be prepared to make liberal use of such tender plants (including verbenas), and they'd also be wise to introduce large groups, or large specimens, of roses. Also, to consider that purple and glaucous-leaved shrubs will make a setting for the red flowers; what Miss Jekyll calls 'using the colour of flowers as precious jewels in a setting of quiet environment, and of suiting the colour of flowering groups to that of the neighbouring foliage'. In the red area of her border she had softening groups of misty white *Gypsophila paniculata*.

There was a 7-ft wide grass path in front of this border, and Miss Jekyll makes the point that a border of this kind should be capable of being looked at from a little way forward if it is to be seen as one picture. If you hug a border too closely you can only look at individual plants.

On the other hand you do want some element of enfilading. Frontally viewed (except at a very considerable distance), all a border's major faults, particularly its gaps, are mercilessly exposed and, again, the eye cannot take it in simultaneously but swivels from side to side.

Miss Jekyll is famed for her practice of bending tall plants forward, as the season advances, so that they cover over those in front of them that flowered early but have gone over. For it must be remembered that, although her border was at its best in August and September, it already included bright spots of colour as early as Oriental poppy and iris time – which is at the turn of May and June. Early colour, however, was never allowed to be at the expense of her ultimate target, the later summer climax. These plants that were bent forward (such as perennial sunflowers) were thereby induced to flower not only at the top of their stems but all along their length. I have never met anyone who has copied this plan of action. My own practice, if I want to blot out, say, delphiniums or alstroemerias once they have become hideous some time in July, is to have something planted *in front* of them that grows up and conceals them or that grows back and covers them. She worked from the back, forwards; I from the front backwards. It would be interesting to know her actual technique for securing her tall rudbeckias and sunflowers and suchlike in an oblique, ground-covering position. She doesn't give it, neither does one gather whether the operation was performed in several steps or all at one go.

I constantly compare my own Long Border (of about the same length but a foot wider) with hers, as also my methods with hers. Although I consider the colours of neighbouring groups of plants, I have no overall colour scheme, simply because this does not appeal to me personally. I include many more shrubs (with roses) than she did, which means that the texture of my border is more varied and so are its contours. I think foliage is possibly a more dominant force with me, although she was strong on this and included rue, santolinas, bergenias, yuccas and stachys in the cooler parts of her border especially. Whether her cannas, in the red section, were purple-leaved, I cannot ascertain. I'm sure she would have liked them to be if such were available.

Although I use annuals and bedding plants, I am not nearly so free with them and I no longer include dahlias, as I once did. This is largely a question of labour, which is more of a worry nowadays than it was in her time. For the same reasons and because I have only one small greenhouse, I do not expressly raise plants in pots for plunging in the borders, but I do sometimes plunge a fuchsia here or a pot of lilies there, in emergency. As for dahlias, quite apart from the staking and tying, I find capsid damage so very obtrusive in a mixed border and that the necessary sprayings are one more thing. I wonder if this pest was as bad in her day.

One practice which I greatly enjoy, when I have the time for it, is the complete replacement of early-flowering perennials, halfway through the season. These early flowerers are lined out in a spare plot and replaced by others – either from pots or boxes in the frameyard or, themselves, from a spare plot. Of the latter, perennial lobelias and sunflowers, monardas and asters all move successfully from the open ground in July and August if properly watered before and after the operation. As far as I know, Miss

Jekyll never did this. I wish I could talk about all these devices with her. That my life only just overlapped with hers so that I met her only once, as a small boy, is a great deprivation.

I include quite a lot of spring bulbs in my Long Border, especially tulips, which she didn't, and one reason I can see for this difference is that I disturb large areas of my border much less often than she did. Also, my soil being heavy suits tulips as permanent features. On her hungry sand they would, very likely, have died out rather quickly. It is the same story with border phloxes. I have large patches of the more brilliant kinds and they are my mainstay for colour in July and August. She found them (and clematis also, which are great water lovers) very tricky. 'They are always difficult here,' she tells, 'unless the season is unusually rainy; in dry summers, even with mulching and watering, I cannot keep them from drying up.' Those she grew, she replanted every autumn. 'The outside pieces are cut off and the woody middle thrown away. It is surprising what a tiny bit of Phlox will make a strong flowering plant in one season.' Yes, but I find that they're far more impressive in their second and third seasons and I usually split and replant after their fourth. Replanting a group annually with single-stemmed units means that you never get the compound group effect; the clumps of flowering panicles within the wider context of the aggregate that composes the complete phlox eiderdown.

Miss Jekyll also replanted her monardas every year and I'm sure for the same reasons. They quickly deteriorate on light soils and need re-establishing.

In studying her border ingredients I feel that there are certain plants I should never want to include. I can see why she did, because of the effect she needed, but the price strikes me as too high. Scarlet salvias and red celosias at the border front. Both plants have aggressively dreary foliage and there's no way of concealing this in so prominent a position. Gladioli were not as large and clumsy in her day as now. The early-flowering *Gladiolus communis* subsp. *byzantinus* and the little Nanus types are easily assimilated when they go over but I should hesitate to bring in the Grandiflorus or even the Butterfly kinds. Not only do they need individual staking and going over daily when in flower to remove the faded blooms and spikes, but they look dreadful immediately they have bloomed. And yet, and yet. Perhaps it could be done by adopting either her or my covering-over tactics or by lifting and replacing them while still green. Gladioli are not so named for nothing. Their proud, sword-like cohorts have a wonderful presence in the mass.

I also envy Gertrude Jekyll's use of tall – that is 4 or 5-ft – antirrhinums. I've never quite had the courage to go all out for these. They need staking, of course. The wonder, you might think, is that they are still offered by the seedsmen, for you never see them in a private garden. The reason is that they are still valuable in the cut flower trade for cropping under glass. And for this same reason you can still acquire single colour strains, from a few seed houses, instead of the ubiquitous mixtures.

Seed strains is another subject on which Gertrude Jekyll elicits cries and grunts of approval from Christopher Lloyd.

'There seems to be a general wish among seed growers just now to dwarf all annual plants,' she observes. While this is useful if the plant's natural habit is diffuse yet 'there seems to me to be a kind of stupidity in inferring from this that all annuals are the better for dwarfing. I take it that the bedding system has had a good deal to do with it. It no doubt enables ignorant gardeners to use a larger variety of plants as senseless colour-masses, but it is obvious that many, if not most, of the plants are individually made much uglier in the process. Take, for example, one of the dwarfest Ageratums: what a silly little dumpy formless, pincushion of a plant it is! And then the dwarfest of the China Asters. Here is a plant (whose chief weakness already lies in a certain over-stiffness) made stiffer and more shapeless still by dwarfing and by cramming with too many petals.'

All this is patently true now as then but it needs repeating again and again. I am always glad when Roy Hay inveighs against dwarf plants and points out, as he fervently does, that, among other disadvantages, they are harder to grow well than those of a more natural stature.

Apropos of ageratums, I remember visiting a trial of this flower at Wisley and being particularly taken by a strain of mixed colours called Mosaique Group. As well as different mauves and blues it included white and charming old rose shades. Furthermore the plants, although all more or less dwarf, were yet of different sizes and heights and this added to the interest of the tapestry. Yet the very reasons for my deriving enjoyment from the mosaic pattern of Mosaique were the same reasons for disqualifying them from any kind of award. Uniformity of dwarfness, of flower size and of other plant features are the standards by which the seedsmen who submit their wares for trial expect them to be judged. I once thought I should enjoy being a judge on the sub-committee that visits these R.H.S. trials but I realize that I should not, for the judges – whatever they think would make the better kind of garden plant – are themselves caught up in the system. They have to judge by those dreary standards of uniformity that are accepted by hybridizers, commercial growers, parks departments and, presumably, the general public.

The grower, Miss Jekyll continues, loses sight of beauty as the final consideration. He doesn't know where to stop. 'Abnormal size, whether greatly above or much below the average, appeals to the vulgar and uneducated eye, and will always command its attention and wonderment.' She concludes this chapter called 'Novelty and Variety' (in *Wood and Garden*) by hoping her good friends in the trade will understand that she is not being personal about them. 'I know that some of them feel much as I do on some of these points, but that in many ways they are helpless, being all bound in a kind of bondage to the general system.'

Growers, hybridizers, judges and public are all in the same straitjacket. If you're in search of freedom and individuality you must not look for it among the majority who run our institutions.

Gertrude Jekyll was remarkable, in fact, for her humility. She never talked down to the ignorant reader or gardener and was ever sympathetic towards their elementary difficulties. But she had no patience with those who wanted to hear their own voices and let loose a flood of questions without listening to or even waiting for the answers. Impatient, too, with those who think it will be easy to copy some pleasing effect but are discouraged the moment they discover that the task is more difficult than it looks.

Arrogance combined with stupidity she found especially hard to bear. As when she had a visitor who owned a rather larger place and thought that, since hardy flowers (hitherto considered as more or less contemptible) were evidently now in fashion, he had better have them. He had lately made a large flower border and speaking of this while they were passing hers, then at its summer best, he said, 'I told my fellow last autumn to get anything he liked, and yet it is perfectly wretched. It is not as if I wanted anything out of the way; I only want a lot of common things like that,' waving a hand airily at her precious border, while scarcely taking the trouble to look at it.

Visitors, however, deserve another chapter.

VISITORS

From year to year, greater numbers of us are opening our gardens to the public. Indeed, for the past fifty years, garden visiting has been a national pastime and recreation, as popular in Scotland as in England.

Initially, we open our gardens for one afternoon, in aid of a charity. It is a tentative experiment; we are convinced that nobody will come. But if the charity is well organized and publicized, come they will. There is a good arrangement, nowadays, within the National Gardens Scheme, which is the oldest gardens-opening organization and the best known, for several small gardens within close range of one another to make themselves open on the same afternoon, allowing one ticket to admit its holder to all of them. Thus the small garden is brought into the net, and this is obviously important because most visitors will themselves own small gardens and they can therefore relate their own circumstances more easily to those they find around them.

Being visited by strangers is, as an occasional experience, enjoyable. Most of them are well-behaved and appreciative and they leave you with a feeling that the effort was worth while. Of course you may be among those who consider admittance to their gardens of the public an invasion. Basically this is because you are selfish and arrogant and look down on them (or would like to feel you can look down on them) as, in various ways, inferior to yourself. 'Thank God that's over!' you exclaim, as you punctually lock the gate a few minutes before time, possibly raising your voice sufficiently to be overheard by the last departing guest.

There's a lot of this raising of the voice to be overheard, in garden visiting, and it more often works the other way. Because there are more visitors than visited, the owners are often likely to hear remarks that are not made directly to themselves, but are meant for them just the same (more on this, anon).

You'll find this, cumulatively, the more you open and the better your garden becomes known. For, having opened but once in the year, originally, there comes a day when the heavens open, it pelts all afternoon and the entire effort is wasted. Clearly the answer is to open on two or three days around the same date and perhaps on another, in spring, to catch the daffodils, with luck.

There it may stop, but it may not. Finding that people rather like coming to your garden and prove it with their feet, you think, 'Why should this all be for charity? Why, additionally, shouldn't we open for ourselves and thereby put aside a few coppers to help pay for the labour and materials implicit in running a garden at all?'

Eventually, some of us find ourselves opening for a great deal of the spring to autumn season. By the end of it, we feel a little jaded; glad of the respite that winter brings, from being constantly on one's toes. We can be untidy again, if we feel like it, not move the barrow out of the middle of the path before 2 p.m., not immediately clear up the rubbish we've pulled out of a border, not mow the lawns as soon as they need it. Bliss.

But when the spring comes round again, we realize that it's been rather monotonously quiet all those months. We've been just a bit too idle and inattentive to the garden's needs. We pull ourselves together and are pleased there's a stimulus to getting ourselves organized and the garden presentable once more.

'We're back again,' we hear from some familiar face, and it really is warming to be appreciated in this, the most effective way, by return visits. Indeed, some of my closest friendships have been built up from a beginning like this; visits repeated, getting talking about our mutual interest and finding that we have much common ground.

If only I were better at recognizing people from one year to the next. I'm not just bad at it, I'm terrible. If I saw them again the next day or even the next week, I should know them (of course). But a whole year or even more – I'm lost. And it can be embarrassing after the tenth time. Almost as though I intended an insult. Perhaps it is insulting to be so forgetful of people one has enjoyed chatting with, but the fact is they are liable to go completely out of my mind until I next see them. They, of course, recognize me because they expect to see me here.

Such visitors don't always make it easy for you. 'Well,' they say, 'You know me,' or 'Who am I?', and are quite disinclined to help. Can you ever deceive people by pretending to recognize them, wreathing your face in welcoming smiles while you gaze confidently at them and start a carefully worded conversation that will at the same time not give your ignorance away but perhaps throw up the necessary clue so that the blessed moment of revelation arrives? I must admit to being a bad actor, on these occasions. I'm so preoccupied with racking my brains, that my unhappy condition is only too apparent.

The less often you open to the public and the fewer your total number of visitors, the greater will be the proportion who are keen and serious gardeners, intent on picking up ideas. If I see someone with a notebook, I'll go and speak to them straight away, in the certainty that it'll be worth being helpful. All too often, alas, visitors will ask the name of a flower, regardless of whether it is labelled, simply for the sake of asking, and I know well that they'll forget it the moment it's been told. 'I'll tell you when I see pencil and paper,' I say, firmly. After all, most plant names are not simple.

Label pickers are a terrible nuisance. I do believe in the importance of a garden that is frequently opened being properly labelled, although I seldom have mine that way. It's an expensive and time-consuming business but it is important, and I have label blitzes in the winter (sometimes masses get written, but there's no time to put them out!). All those I use are of the stick-into-the-ground type, as mine is not, essentially, a tree-and-shrub garden. Visitors, who mostly have bad eyesight, will pick them out whether of the ground or of a pot. That's fair enough, but will they stick them back again properly? Almost never. I do implore the reader, if he or she is one of those I have in mind, to note, when pulling the label out, exactly where it came from, and to push it firmly back when read. Labels are originally set in their positions when the soil is damp and they go in easily, but often in summer the ground sets like concrete (anyway on clay soils), and unless the label is returned to its original slot it won't go in at all. But that's only half the story, because most visitors lose all interest in the label's fate the moment it has been read, and it won't be pushed into the soil, however soft. It won't even go back into the same pot or in front of the same garden plant because by then her (or his) attention has moved on to something else or she's talking volubly the while, to a companion.

Thus do the labels get scattered all over the place and this is almost as disheartening as when they find their way into the occasional dishonest visitor's pocket or bag (really durable, expensive labels are out of the question nowadays, for this reason).

Coachloads of garden visitors are less likely to be interested in the plants than in living through the hiatus before tea time. Their standards of flower preferences may tend towards the brilliant, to an extent which yours may not always satisfy. I enjoyed the description of the phloxes in my Long Border by one woman who was evoking the scene to a blind friend: 'Oh Gracie, it's a picture; every shade of pink and mauve!' But when the phloxes are over? I remember another visitor in October exclaiming, 'Of course, all the colour's gone.' Nettled, I suggested that if she removed her dark glasses and stood with her back to the sun instead of facing it, she might be able to enjoy the fuchsias and a few other oddments (I'm rather fond of my garden in October, as it happens). But I might as well have retained a stony silence as she now did. Silences speak volumes. 'I gave her a look,' one friend was wont to relate, when describing some foolish or offensive remarks that had been addressed to her. Much the best way, rather than trying to score points. I doubt if I shall learn it.

A feature in our garden that upsets the type of visitor who regards orderliness and civilization as synonymous, are the areas of grass that we allow to grow long and to flower before we cut them. They are like an old-fashioned meadow before weedkillers, artificial fertilizers and the ley system of farming became ubiquitous. I have talked about the attractions of such meadow gardening sufficiently elsewhere, but the basic point at issue here is that, while it may attract you, it may, on the contrary, strike you as repulsive.

The first thing visitors see, as they lean over our front gate either wondering whether it's worth coming in or hoping to see as much garden as they can without the slightest intention of coming in, is two large areas on either side of the front path, of very rough grass (the season being May, June or July), enclosed by a formal outline of neatly trimmed yew hedges. They can hardly believe their eyes. 'Your lawns are weedy,' one such visitor accused me (we actually use weedkillers fairly effectively on our mown lawns), and 'It's a pity about the weeds,' was another's comment, not meaning the cultivated areas but the flowers in the grass, while a third exclaimed, 'You *do* grow parsley,' indicating the cow parsley. 'We like it,' I retorted and he gave a snort. Some, speaking more in sorrow than in anger, commiserate with me on how difficult it must be to find the labour nowadays.

On the other hand some visitors, delighted at recognizing a wild orchid, plunge without ceremony into the deep field, to get a close look. You cannot walk in long grass without leaving a very ugly track of squashed vegetation. This does not occur to the impulsive. The track having been made, others follow. Darters among the public are quite a menace. For this reason I seldom, nowadays, stick a label in a border that cannot easily be reached from a path. If out of reach, impetuous visitors dart in with eyes on quarry and quite unmindful of the havoc their feet are wreaking.

Photographers can be equally inconsiderate. I'm sorry if I seem to be piling on the agony; it all comes out as I think back. As a reluctant photographer myself, when visiting other people's gardens, I know how tempting it is to place a foot in a border so as to get a closer picture of a plant. Photographers will all too frequently put the interests of their picture first, regardless of where they go in its pursuit. Small wonder that tripods are not allowed into the Savill Gardens at Windsor. Others may follow suit. Yet, curiously, we have never been subjected to the slightest discourtesy from professional photographers visiting us by appointment. They are apparently able to obtain the best in results with the minimum of obtrusion.

Rarely, nowadays, do those opening to the public allow dogs into their gardens. In a park-like, landscape garden, perhaps they won't matter, though even in Sheffield Park they are not permitted, nowadays. At the National Pinetum, Bedgebury, Kent, everyone takes their dogs and it is all great fun; you can learn your breeds, so many are there, and there are squirrels to chase, moles to dig for, all sorts of excitements.

In the normal enclosed garden, excitements are destructive. The owner's dogs will have to be shut up if others are allowed in, else there will be incidents, which always seem to start against flower borders. So, when the organizer of a coach trip has written that 'one of our members has a wee lap dog', it will be wiser to be firm. Even a wee lap dog is all too often set on its four feet as soon as the coast seems clear, and not even on a lead, like as not.

Another case for firmness is with regard to times of opening. If you are remote and only open once a year anyway, you may well take a kindly and hospitable view of the few visitors who beg to be allowed round at other times. But if you make a practice of opening anyway, then out-of-hours visitors quickly become a pest. They will always have a special case to plead, but they couldn't give you notice, not even a phone call to ask if it is convenient. Enough is enough: if you are open several afternoons a week you'll want to be firmly closed at other times, otherwise you'll never be able to work or relax in your garden without interference. If you let some in, the word will get round; you'll have to let in others; there'll be no peace. Of course if you want to be distracted, that's another matter.

I don't think it's a matter personal to me, but a pretty general experience among garden owners, that it is impossible to work in the garden for any length of time while there are visitors around. Inevitably you become an eavesdropper, whether you've been seen or whether you've not. If you've not been seen because you're down on your hands and knees, weeding among tall plants, it can be amusing, but it's embarrassing for all concerned when you're obliged to move or sneeze and reveal yourself. If you have been seen there's a terrible temptation in visitors to raise their voices, ostensibly still speaking to each other but actually with the intention that you shall overhear. At what point are you expected to join in? At what point do you join in, if at all?

'I wish I knew the name of that,' said a woman to her husband. 'Of what?' 'Of that plant there.' 'Why don't you ask the man? I daresay he'd tell you.' 'I daresay he wouldn't.' She glared at me and I glared back, and that was that. Next minute another woman came up perfectly naturally to ask about a plant nearby and it was a pleasure to tell her.

'Do you find you lose much?' is a question I'm frequently asked. Not really. It gets a little worse from year to year but I'm generally as quick at forgetting my losses as I am my visitors' names and faces. Losses can be so annoying as to be unforgettable – for a time. On an important corner I planted a fairly rare plant that I'd been delighted to find just one of in a nursery that was closing down. It was the gaily variegated form of a Virginia creeper, *Ampelopsis brevipedunculata* 'Elegans'. As it is a weak grower my intention was to grow it on the flat, not as a climber. It was stolen almost as soon as it came into leaf, although I hadn't labelled it. I never, nowadays, label a valuable plant until it is securely established. When we replaced it with a rather special yucca, Romke van de Kaa, who came to me from Wisley, fixed it as they have to plants in the rock garden there, with two long wire staples plunged on either side and joined over the top of its rhizome. Nothing shows, but when the thief starts pulling he is frustrated.

Private gardens are not usually marauded in the ruthless and deliberate manner that public and society's gardens are. The very word National in National Trust properties seems to be a challenge. 'If it's national it's mine,' goes the reasoning, and so go the plants.

If a cutting is taken without spoiling the look of a plant and I don't know about it there's nothing to lose sleep over. Too often visitors break and tear cuttings off from the most prominent part of a bush, leaving ugly snags. This is thoughtless. I have a 'Dollar Princess' fuchsia growing on a tempting corner that suffers much in this way. When I go to it for my own cuttings, half have been removed. I should place my plants more carefully.

Finger blight this disease is known as. When it happens to hardy cyclamen tubers it really is maddening, and I am told that enthusiasts for alpine plants are the most dishonest of all. They come equipped, and you may have to police collections of small and precious plants.

A lavender bush overhanging a path is sure to be made hideous by having its flower heads pulled off. Poppy seed heads are irresistible, it seems. I leave my opium poppies for ornament, but those within reach of any path quickly become headless. Lily capsules likewise, and they are usually taken green before they are any use. A label, 'These seeds are being saved', helps against the merely thoughtless. I have an *Abies koreana*, the whole point of which is that it makes its fascinating candle-like cones on plants that are still only a few feet high. Ours produced its first in the spring following the great 1976 summer. Romke and I had both been watching its developing buds, hoping that one at least might produce what we longed to see. But you have guessed. It went, even as early as April, no sooner were we open. I don't know how to stop that.

To raise a bit of extra cash, either for charity or for themselves, many garden openers these days have a plant stall. If you're doing it as a business, some sort of official pricing is obligatory, but if not, then to write names and prices for each plant would be a bore. Much easier to quote prices to prospective customers as you go along. Never hesitate when you're asked. Never put a note of inquiry in your own voice, as though you're suggesting a price subject to the customer's acquiescence. This makes him uncomfortable and suspicious. Worst of all to say, 'It's 90p I'm afraid,' deprecatingly, as though you were ashamed of asking so much. Think ahead, sum your customer up for what he's likely to be worth before he puts his question, name your price firmly with a brief candid glance at him (blue eyes are a great help), and then talk interestingly about the merits, rarity and cultivation of the plant in question until the deal is concluded.

WITH FARRER FOR REFERENCE

In his life, Reginald Farrer cannot have been an easy companion; exceedingly moody, he was often either on top of the world or in the depths of depression and he had, besides, the unfortunate physical disability of a hare lip and cleft palate – enough to twist anyone's character. But in his writings he was at his best and his alphabetical work of reference, *The English Rock Garden*, is a masterpiece and classic. Particularly when you consider that it was first published in 1919. Similar dictionaries and encyclopaedias on gardening had, up till his time, been exceedingly stodgy, not to say indigestible. Useful they might be, but not a pleasure to read and only to be taken in very small doses.

William Robinson began to break away from this opaque tradition in *The English Flower Garden*, but Farrer carried it much further. He was a compulsive writer and he was devoted to plants in a most personal way. It was thus a labour of love to convey his knowledge and his sentiments to a hungry public. For those who really do desire to know their plants, he said, there is no reason why the necessary details 'if clearly and simply conveyed, should not only be helpful, but even readable and illuminating'. We take such an approach for granted today, but he was breaking new ground.

He had a great command of language, a visual and analytical memory for the plants he was describing, and he had wit and a delightful flare not only for eulogizing but for debunking. Again and again he strikes me as the equivalent in his field to Donald Francis Tovey in the realm of musical analysis.

I am neither an alpine nor a rock gardener and so, when these plants of specialized habitat impinge themselves on my wondering and even alarmed consciousness, I turn to Farrer. Take androsaces, for instance. I don't grow one of them. We used to have good colonies of *Androsace studiosorum* (syn. *A. primuloides*) and *A. lanuginosa* at Dixter, but they have gone and not been replaced. When a monograph on them appeared in 1977 I read it from a self-centred viewpoint wondering how much I was missing, how much I might adapt without needing to alter the style of gardening that suits the place where I happen to live.

With a sigh of regret I came to the conclusion that, given the damps of our climate and the depredations by birds of rosette-forming plants grown outside, androsaces were not for me. If you are prepared to grow them in pots and move them under glass in the winter, that is another matter. And it was a relief, on turning to Farrer, to find him admitting (what the authors of the monograph never suggested) that 'of all mountain-races this name is engraved most deeply on the rock-gardener's heart, like Calais on Queen Mary's, standing as well for his highest hope and pride as for his bitterest disappointments.'

Sometimes I turn to him in the wake of an Alpine Garden Society show, when this coincides at Westminster with one of the R.H.S.'s. Such beautifully grown plants are seen there and I make a good many notes, finding that this is the only way to fix in my mind's eye the look of a plant I shall not see again for a long while. If you force yourself to describe it in a note you force yourself to look at it properly, analytically. Thereafter you must re-read your notes within twenty-four hours and then again, within the next few days, do a bit of looking up. Unfortunately Farrer is not alive today and many plants we see at A.G.S. and other shows are not to be found in his great work, but others that you would not have expected to be included, are.

For instance, *Viola pedata*, which you may see at a spring show. As any layman like myself can tell, the genus *Viola* divides itself horticulturally into two distinct sections: that of the violets and that of the violas and pansies. A violet is always easily recognized as a violet even when, as in this case, it assumes a partial disguise by wearing a pedate instead of the usual ace-of-spades-shaped leaf.

Not many garden plants have pedate leaves but those that do are all exceedingly stylish. *Dracunculus vulgaris* (the dragon arum), *Helleborus foetidus* (the stinking hellebore) and *Adiantum pedatum* (the most arresting of the hardy maidenhair ferns) are those that spring to my mind. The leaflets radiate from an arc-shaped foundation.

On the genus *Viola* Farrer tells us that it 'brings this alphabet to the last great dragon in its path. No race is more fertile of more exquisite beauty, but no race is also more fertile in dull and dowdy species. And unfortunately, in these later years, the enormous multitudes of American violets have taken (no less than American heiresses) to overflowing into our continent undescribed…'

Viola pedata, from the sandy levels of Long Island, is well described but has, it seems, always been a problem to the cultivator. It certainly has that look about it. 'On the whole it seems likely that a sandy and perfectly-drained woodland mixture, light and free, but specially rich and clammy and well watered from below in spring will best appease the plant's homesickness, if only it be enshrined in some place not too open to the furies of the sun.'

Is not clammy good there? Indeed, the whole lyrical recipe seems bound, if followed, to appease the most captious and irritable Princess on a Pea. 'It is well-deserving of all the trouble it gives, so delicately beautiful are the tufts of quite dwarf and dark bird-clawed foliage; and so splendid the noble orange-pointilled ample violets with back-turning upper petals, hovering on their 2-to-3 inch stalks like golden-bodied butterflies of a lucent blue lavender.'

Some readers, I know, think this kind of torrent is uncomfortably over-written, but others, and especially the enthusiastic plantsmen among us, revel in Farrer's evocative range that so successfully conveys his enthusiasm and echoes (or is echoed by) our own. The stunted imaginations of the majority of gardening writers are only too familiar and abundant.

There is another gem in *Lithospermum oleifolium*, with distinctive obovate tear-drop leaves and small clusters of pale blue flowers on a plant of open yet not sprawling habit. 'A dwarf diffuse mass of shoots', Farrer calls them, and of the flowers, few and rather large, 'of a specially lovely opalescent violet, with shifting lights of blue and pink to deep and pale. Those who have grown it well shall advise upon its culture. It is not a thing as easy to meet with as dullness or folly, nor as easy to keep as a plain daughter.'

In the same borage family, *Omphalodes luciliae* has an unmistakably delicate expression, with its glaucous foliage and flowers that I merely described in my note as pale blue. But hear Farrer, who knew it in its native 'god-haunted mountains of Asia Minor, delighting those inhospitable rocks with abundant loose sprays of its round blank-looking flat flowers in the most delicate pearly tones of very pale porcelain-blue or sun-kissed snowy rose, seeming there to develop like natural emanations from the blueness of the foliage'.

To those who complain of Farrer's verbal diarrhoea I would point out that the three stark monosyllabic adjectives 'round', 'blank' and 'flat' as here inserted are the kernel of his description. If you have seen the flower, even on a bench at Westminster, you are immediately struck by their rightness. Yet, simple as they are, would you have used them yourself, in this context? Improbably.

The English Rock Garden is a joy to dip into because you keep discovering gems of description and comment that you had missed hitherto.

UNEASY STATUS

That we should change our minds about the status and merit of the plants we live with is surely no disgrace. At any rate I hope not, because I often find my opinions to be as unstable as a weathercock.

Elaeagnus pungens ' Maculata' used to be one of my favourite shrubs. Such glorious variegation, such luminous colouring in winter sunshine. And yet my doubts and reservations kept increasing, and I had more and more difficulty in choking them back. Then came my let-out. John Treasure was with me on one of his rare visits, and as we passed the elaeagnus he remarked, 'You know, I'm getting *tired* of that shrub.' As his voice is naturally drawling, the word 'tired' evoked Atlas with the entire world burden on his shoulders. Endorsement on a doubtful issue is a great stimulus, and I leapt enthusiastically to the shrub's offence. It doesn't know how to grow; its angular branch system is like a forest of ingrowing toenails, and it is always liable to die back at the tips. Often it just dies: full stop. The solution is to grow *Ilex* 'Golden King' instead. After twenty-five years of its company I still say that this holly has every virtue. True it is slow-growing, but even that becomes a virtue after twenty-five years.

Now what of *Buddleja fallowiana* 'Alba'? I gave this over-praised shrub a shrewd dismissive backward cow-kick some time ago, and that should have settled its hash. I was only waiting to remove my specimen until the shrub behind it (a privet – I have a weakness for privets) had grown sufficiently to minimize the inevitable gap. At that moment the doomed buddleia chose to excel itself. It is still true, as I have often snidely remarked, that the flowers on each spike are not half of them opened before the first are already brown and withered, but somehow, I now realize, this does not matter. The spikelets are slender and unobtrusive whether living or dead. They form, in a healthy specimen, a harmonious unit with the leaves, which are of even greater importance; a tall shimmering dome of different shades of cream and grey.

What of that disputatious plant *Persicaria affinis* (syn. *Polygonum affine*) 'Donald Lowndes'? I loathe it. It was a miserable failure in my Long Border, and I shall never, I hope, attempt to grow it again. And yet how can I help but admire it as seen in some (not all) gardens: a low ground-coverer with a forest of thick spikes whose colour changes from rose red to brick red and finally, with the foliage, to warm brown as the season advances? How can it be so handsome and effective and yet, basically, such a horrid plant? Well, there are people like that, too.

P.a. 'Donald Lowndes' is not just coarse; it is coarse-grained to its innermost fibre. 'Much finer flowers than the above, good ground cover,' Bloom's catalogue tells us. 'The above' is 'Darjeeling Red'. Now it all depends on what you mean by finer. If you mean larger and having more flower power, agreed. But if you mean having greater refinement, then 'Darjeeling Red' (deep pink, really) has it every time since its spikes are slenderer and its leaves don't stare at you with bold unwinking insolence.

I developed a rapid and full-blooded hate for the hybrid hosta called 'Honeybells'. Having *H. plantaginea* as one parent it retains that species' night scent but adds pale lilac colouring to the dead-whiteness of *H. plantaginea*'s flower. So much for the gain (if gain it is); the losses are disastrous. 'Honeybells' has lost most of the pristine freshness so notable throughout the season in its parent's foliage. It is, moreover, far too vigorous. After two years my colony was already so thick and congested that flower spikes were produced only at the margins. Other gardeners tell me of the same experience. Furthermore, if there is the slightest setback in growing conditions at the time of flowering – I mean through shortage of water – the buds drop off the stems unopened. There was a wonderful rending of roots and then a beautiful space for re-planting. I am told that I should prefer 'Royal Standard' but there are already too many hostas around.

Plants do, of course, behave differently for different people. An acquaintance whose garden I was visiting for the first time led me purposefully to her specimen of wintersweet, *Chimonanthus praecox*, growing against a white wall. It was still in full leaf. I have written in the past of this shrub's 'sordid appearance in its summer dress', for the leaves are large and coarse, and yet, in the instance I am describing, the leaves were large but not coarse. They held themselves with notable distinction. Very strange. On returning home I rushed out to my specimen with the same predisposition for indulgence as the father showed his prodigal son, but it was no use. My wintersweet was as scruffy and unprepossessing as an unshaven tramp. As usual there was nothing for it but to wait upon leaf-fall as a merciful release.

PRICKLES AND SPINES

If every cry of pain emitted by the unwary gardener as he impales himself on the sharp point of a plant could be collected as it was made all over the world, it would surely add up to one unbroken ululation, the sound varying according to language between the narrow limits of Ow! and A-eee!

Why do we grow such plants? Well, in the first place there are those among us who are in a position to command but who never themselves dirty their hands. Thus insulated, it is easy for them to say, 'Just prune the 'Mermaid' and 'Albertine' roses, will you Bill, and tie in all those nice long young shoots.' Note the throw-away use of the words 'just' and 'nice', suggesting first that the job can be tossed off effortlessly, painlessly and in a matter of minutes; second that the niceness of the young shoots is somehow extended from their indisputable youth and vigour, so as to embrace their temperament. Bill will very soon find out, if he didn't know already, that as soon as one of these nice young shoots, still waving about with the freedom of an untamed savage, is approached, it will grab him in the back with its handsome, fresh brown hooked thorns. And not merely in the back, from which single assault the hapless victim might free himself with a downwards and sideways convulsion, but simultaneously in the back of his scalp also. In the circumstances, leather gloves must be accounted inadequate protection. Nothing less than a full suit of armour, cap-a-pe, would save. And when it comes to the tying in, who can fiddle about with knots wearing leather gloves? Away they have to go. If the weather is cold you can contemplate your bleeding hands with a certain detachment. 'Look at all that red stuff spilling around and getting on my clothes,' you muse, 'where does it come from?' The pain is felt only when your circulation returns to normal.

It may be thought that the lordly dispenser of unpleasant duties is a rare bird these days, when most gardeners are amateurs and have to do a job themselves if it is to be done at all. But this overlooks our vast public gardening sector, wherein curators, superintendents and supervisors abound. True, there may be more rhododendrons than rambling roses within their realm.

These overlords will also be more keenly aware than we need to be of the uses to which armed plants can be put as offensive weapons. Plants in public places are easily destroyed by unheeding, insensitive people who simply trample on them when it suits their convenience. 'This used to be a short cut. It's not a short cut any more,' was the laconic comment made to me by the superintendent of the gardens surrounding a university's halls of residence. He had planted the cut in question with bushes of the sea buckthorn, which is a good deal more than just pretty.

Both reasons so far adumbrated for growing vicious plants are based on the remoteness of their nasty natures from the person who decides to grow them. When we bring the gardener himself forward as arbiter of decisions, there are some plants whose beauty consists in their very spininess and this, to him, may make their cultivation irresistible. I don't think I belong to this dedicated community. Walking round my garden with two friends, one of them stopped in front of my cardoons and said he must give me seed from his, which was even pricklier. I must have let out a small involuntary moan because he then turned to our companion and said that he would send them to her instead, as they would make superlative material for her flower arrangements. I breathed again.

Friends who think I ought to grow it constantly ply me with seeds of the thistly *Onopordum acanthium*. It is, I admit, as handsome a brute as you could wish to clap eyes on; 8 ft of it all silvery but clad from top to toe in spines and prickles pointing in every direction. Half-way through the summer the plant, as it runs to seed, suddenly turns dowdy and tarnished and requires removing. I know only too well that however carefully I rehearse a throw-away manner in giving the order, 'Just clear away the onopordums, will you B...?', when it comes to it, I shall be clearing them away myself. I'd rather not grow them, thank you very much.

Wounding techniques by plants make an interesting study. When Winnie the Pooh had fallen into a gorse bush, he emerged (according to both the text and the illustration) covered with prickles. But I consider this most unlikely; that would have been the technique of a prickly pear or any other *Opuntia*, but not of gorse. In a comparable way, a bee (alias prickly pear) leaves its sting behind for us to remove as best we may whereas the wasp (alias gorse) retains its apparatus. Once free of the wasp we are free of its sting. If only wasps made honey I should prefer them to bees; they're not nearly as treacherous or bad-tempered.

Further parallels can usefully be drawn from dog bites. Some dogs have a cleaner biting technique than others. While one will leave a messy gash, another (with a snipy jaw) will go in and out so neatly that a nylon stocking will remain undamaged, though the flesh underneath has been punctured. What's called going straight to the point.

The curved thorns on roses are the messy, dirty biters of the plant kingdom. And as to treachery, what shall we say of 'Zéphirine Drouhin' which claims to be the rose without a thorn, but keeps a few tucked up her legs just to catch us out? The nastiest jab of all is inflicted on the unsuspecting hand that moves confidently and without aforethought in areas where no opposition is anticipated.

By contrast, the more belligerent yuccas, as also the New Zealand spear grasses (*Aciphylla*) and gorse itself are typical of the plant from which you do really know just what to expect. A sharp stab is your punishment for taking liberties, but then it's all over and one must hope that the damage inflicted was not such as Fanny Squeers described when Nicholas Nickleby assaulted her ma and with dreadful violence 'dashed her to the earth, and drove her back comb several inches into her head. A very little more and it must have entered her skull.' Careful and selective as I may try to be, my garden is still littered with back combs.

THESE FLOWER NON-STOP

Of particular value to the preoccupied gardener are those perennials and shrubs that have the capacity for flowering for three months at a stretch. Many annuals and tender bedding plants are grown for this express reason, but their tenderness or short life commits us to quite a bit of extra work, which may be resented.

However I must admit about some of the plants I am rounding up that they may be short-lived if left to get on with it for very long or, alternatively, they may flower very much more freely and over a greatly extended season if you will take the trouble to lift, split and replant them in fortified soil every year or two. But I'll bring in these points as I come to them.

At the lowest level and positively requiring no sort of attention is the ivy-leaved toadflax, *Cymbalaria muralis*. Generally colonizing dry or rotten walls or the cracks in paving, it carries small mauve snapdragon flowers in mild seasons the year round, but more usually from March to December. In the wrong place it can become a weed, as when it starts clambering over low box hedging, but it couldn't be easier to control. There is an equally likeable albino, and as is often the case with true albinos, the leaves are a noticeably bright, fresh shade of green.

Erigeron karvinskianus (syn. *E. mucronatus*) colonizes just the same sort of places but is a daisy. Originally from Mexico, you see it growing in a feral state in many parts of Europe including Cornwall and the Channel Isles, but it is not hardy enough to survive in our cold Midlands, for instance. Here, in Sussex, I lost plants in the excessively severe winter of 1962–3, but it regenerates so freely from seedlings that the losses were made good within months. *E. karvinskianus* opens its first white daisies, which turn pink as they age, in May and continues thereafter with unabated zest until the first severe frosts of winter. In a frostless winter such as we experienced in 1974–5, it flowers right through.

Another kind of erigeron, *E. glaucus*, has a first glorious spate of blossom starting in May and then continues at an agreeable jogtrot till the autumn.

This species is particularly common in seaside gardens but you rarely see it listed. Its flowers are substantial, borne in trusses on a mat-forming plant, and they are some shade of mauve or pinky mauve. The cultivars 'Elsie', 'Ernest Ladhams' and 'Elstead Pink' belong here.

Rhodanthemum (syn. *Leucanthemum*) *hosmariense* also carries sizeable daisies, white in this case, but on a more distinguished-looking plant making a reasonably tight hummock of finely divided steely grey foliage. Even when only a few flowers are present, they make their impact. *Osteospermum jucundum* (syn. *Dimorphotheca barberae*) is one of the hardiest of its tribe and flowers at all seasons till the new year, but is freest in June. It makes a large, loose, space-consuming mat and seems as happy in shade as in sun. A South African-style daisy, it is bright mauve pink with a yellow disc. *O.* 'White Pim', on the other hand, is white (blue-mauve on the reverse) with a blue disc. It has a sprawling habit and is a good tub plant. Since tubs are apt to be rather dry in winter (when no one thinks of watering them) and exceedingly well drained, and as they usually occupy sheltered positions, this is the spot where you are most likely to bring 'White Pim' through the greatest number of winters. But it is wise to safeguard yourself with a few cuttings in a cold frame each autumn.

Cuphea cyanea also benefits from competitive starvation treatment as generally received in a tub. Related to the Mexican cigar plant, *C. ignea*, it is difficult to believe in its remarkable hardiness – much hardier than the dimorphotheca. It grows 18 in. to 2 ft tall and carries a long succession of small tubular orange flowers, yellow at the mouth, from late summer till November. In well-fed border soil the colouring is wan, but in a tub it is bright and shows off to effect in association with orange and bronze *Mimulus glutinosus* and *M. puniceus*, all intertwined with the grey foliage of *Helichrysum petiolare*. I was proud when this cuphea received an Award of Merit on my showing it to the R.H.S. on 31 October 1978. (In fact I felt that the award had been made to me.)

Oenothera macrocarpa (syn. *O. missouriensis*) is a day-and-night-flowering evening primrose (without scent, however), carrying large yellow goblets on a hardy prostrate plant. Individuals look thin and stringy so it is worth grouping this one for a carpet effect. Season: June to November. Most oenotheras, in fact, are long-flowering. So is their close relation, *Epilobium glabellum*; 9 in. tall, its creamy flowers are borne with astonishing freedom from May till autumn. There is a price to pay: the plant is apt to die out, which comes as the greater surprise in a member of the invasive willowherb tribe. Young plants are free in their production of short, leafy shoots and these should be made into cuttings at any time between autumn and spring. Ventilated conditions are necessary; they damp off if kept close.

I have often written of that paragon, *Viola cornuta* Alba Group. Moisture is its one demand and it will then flower from May till November, rambling through its neighbours. Possessed also of this invaluable weaving habit, which helps integrate a border's ingredients, are a number of hardy cranesbills. *Geranium sanguineum*, the bloody cranesbill, has blue blood of a vivid magenta, generally speaking, though it is variable in this respect, and I have also a bright pink cultivar obtained years ago from Jack Drake's nursery. There is an albino and also the pink-striped variety *lancastriense*. The main spate of blossom starts in May and I cut my plants back in late July, whereupon they crop again in the autumn.

Geranium (Cinereum Group) 'Ballerina' does not ramble. I grow it at the front of my borders but really haven't the right place for this very low but steadfastly flowering plant. Against a pale mauve background the petal veins and flower centre are picked out in purple.

Geranium endressii is among the most valuable shade-tolerating stalwarts. Its mauve-pink colouring is very slightly abrasive but not really so in such a small flower and in shade. Anyway I prefer this, which is self-supporting at a foot or 18 in., to its more vigorous hybrids, salmony *G. × oxonianum* 'Wargrave' and the rudely vigorous *G. × o.* 'A. T. Johnson', a paler, silvery pink. The latter is apt to swamp its neighbours but will climb endearingly to 4 or 5 ft into a near-by shrub. All these flower non-stop from May till autumn. The most generous of the lot in this respect is 'Russell Prichard', which will make a pool of magenta on paving in front of a border's margin while climbing into taller plants behind it; good with the pure white *Phlox* 'Mia Ruys', for instance.

Geranium wallichianum 'Buxton's Variety' is generally called 'Buxton's Blue' but as it is invariably raised from seed, has no claim to a clonal name. It flowers from July till late autumn and early blooms are apt to be a disappointingly muddy pink-mauve. This is a reaction to hot weather. As the year cools off (and this is an excellent cranesbill in shade) the colour improves progressively to near blue with a white eye and purple anthers, the leaves being interestingly shaped and mottled. It is a rambler and must be given plenty of elbow space. I suggest interplanting it with colchicums. These will have done their leafing in spring before the cranesbill, which is a late starter, has need to spread.

Convolvulus sabatius (syn. *mauritanicus*) is a short distance trailer and must not be allowed to be crowded out. Its funnel flowers are an innocent shade of campanula blue and open in succession from May till late autumn. Leave its old growth as a protection through the winter and early spring and, given a well-drained, protected site, you are unlikely to lose it.

The mat-forming *Persicaria affinis* (syn. *Polygonum affine*) 'Donald Lowndes' is not my favourite plant, but it has to be conceded that a number of persicarias have a long season and, when suited, this is one of the more effective. We didn't suit each other. At the 3-ft level, *P. amplexicaulis* 'Atrosanguinea' is a thoroughly sturdy and reliable plant carrying crimson pokers in abundance from July onwards. I recommend it to *you*. And there is *P. milletii* with sorrel leaves and a long succession of the most vivid crimson clubs I've seen in any knotweed, but they are a bit too heavy for the strength of their 18-in. stems and there are never quite enough of them at one time. I think if you grew this on very rich moist soil and replanted it frequently it might be spectacular.

Frequent replanting is the prerequisite for the pure white double Shasta daisy, *Leucanthemum* × *superbum* 'Esther Read'. It is so generous with its blossom as to come into the flowers-itself-to-death category unless re-set each spring. Only 2 ft high, its flowers are anemone-centred – ideal for wreath work, but as wreaths are now rather out of fashion perhaps they could do duty in daisy chains.

'Esther Read' is a good border plant but *Crinum* × *powellii* should mainly be considered as a cut flower, since its foliage is outrageously tumbled, like an unmade bed. But this strong-growing member of the amaryllis tribe with its heads of pink or white funnels is amazingly hardy and flowers from July till November.

Neither shall I recommend *Scabiosa caucasica*, of the large blue pincushion flowers, as a border plant, because it never does enough for you at one time to make a real display, but on light soils and in a sunny position I should include *S. graminifolia*. Quite apart from its long succession of mauve flowers, its narrow, silvery foliage is always a picture. Gaillardias I dismiss from my heavy soil, where their habit is straggling and weedy and they are anyway short-lived and a prey to slugs, but on light soils they are good value.

Because of the carnation blood that runs in them, Allwoodii pinks give tremendous value over a long season. 'Doris' is the best known; I cannot stomach that simpering salmon shade. These pinks become scrawny after three years of hard labour and must be replaced.

Euphorbia seguierana subsp. *niciciana* is undoubtedly the longest flowering of the spurges and a bright lime green on 18-in. mounds from May to October, but not long lived. Self-sown seedlings turn up, but you should make certain of your stock by taking cuttings of young shoots (preferably from young plants) at any time in the summer.

Verbena bonariensis and *V. rigida* (*V. venosa*) might be described as the long and the short of it. The former grows to 5 or 6 ft but is a see-through plant, more stem than leaf, and you can place it at the front of a border (unless you lose your nerve). The latter is a favourite with public parks and gardens as a bedder, for its colour is a brilliant shade of purple. It is a reasonably hardy perennial, in fact, and only 2 ft tall. Both flower from July till November and are usually raised from seed, which germinates slowly.

Japanese anemones (*Anemone × hybrida*, syn. *A. japonica*), in pink or white, flower from early August till mid-October; *Aster × frikartii* starts at the same time and lasts almost as long – an excellent lavender shade and self-supporting once established. *Rudbeckia fulgida* var. *sullivantii* 'Goldsturm' has the same season and, given moisture, is admirable for lighting up dark places.

When we turn our attention to long-flowering shrubs we find that most are a little bit tender and need to be near a warm wall. Or else, like *Lavatera × clementii*, they are short-lived. On my heavy soil I'm doing well to get three years from the tree mallow but on sand or chalk you might get five. Anyway, cuttings strike easily at any season. You should search out a good strain of this valuable plant as its flowers can be miserably small or they can be large and handsome and always in a pleasing shade of mallow pink. It grows 5 to 6 ft in as many months and flowers from June on and onwards. The hardier penstemons should give you three or four years, but it must be admitted that such as 'Evelyn', 'Andenken an Friedrich Hahn (syn. 'Garnet'), 'Drinkstone Red' and 'Sour Grapes' flower most freely and continuously in their first year. Strike cuttings in August and plant them out in April. Subsequently old plants should be cut very hard back, almost to the ground, each spring. They will then make a fantastic display in June–July but will yield only a sprinkling thereafter.

Phygelius differs from the penstemon mainly in coming from the Cape instead of from the New World. *P. capensis* is the hardiest but will still flower most freely near a warm wall, with large panicles of its terracotta, tubular flowers. It blends ideally with the feathery grey foliage of *Artemisia arborescens*. The less well-known *P. aequalis* is a soft and subtle shade of dusky pink that I even prefer. It now has a primrose yellow form.

Among the most rewarding shrubs for a warm corner is *Salvia microphylla* var. *microphylla* (syns. *S. grahamii*, *S. microphylla* var. *neurepia*). It seldom grows more than 3 ft by 3 ft and carries brilliant carmine flowers – sometimes nearer to purple. You have to tidy it up and remove dead wood each spring, but if it has brought sufficient through the winter alive, its season starts in May and continues into late autumn. Makeshift though its structure is, it is a remarkably long-lived sage.

Abutilon megapotamicum must be trained against a wall. Its twigs are slender and lax but it has to retain a good framework of these in order to cover a large area. In the spring you can trim it a little. To say that this abutilon will flower from July to January is to miss the point. It must not just flower but flower abundantly – huge swags, trails, ropes of its small red and yellow lanterns with their protruding brush of brown stamens. Otherwise it might just as well not be there for all the effect it contributes. Its leaves, though elegant and small (for an abutilon) are a particularly dark and gloomy shade of green. When flowering with abandon, however, the leaves virtually disappear. This wonderful condition – and it is wonderful – can never be predicted and seldom occurs two years running but it is well worth planting and working for.

On the other hand I have the feeling that the common passion flower, *Passiflora caerulea*, never brings it off because it never produces its goods in sufficient quantity on any one day. Each flower only lasts a day, so the chances of having a hundred or more out simultaneously are remote. A dozen or twenty is the usual form. But, if you have a free-flowering strain (many are not), this is undoubtedly a warm-wall plant with a long season and its flower structure is unique. The plant itself, alas, is structureless and can look quite a mess.

To my chagrin I have yet to score a success with *Solanum laxum* (syn. *S. jasminoides*) 'Album', although it is a *succès fou* in London. It is a true climber of the potato variety with loose clusters of pure white, jasmine-like flowers, each with a yellow eye of stamens. If it brings its old wood through the winter, flowering starts in June and continues till November. It is far more spectacular than jasmine itself, though without the scent. Given reasonably mild winters, this shrub will even flourish in Scotland, near the sea. I've seen a splendid example at Aberlady, in East Lothian. Yet it fails for Lloyd. I can flower it in a pot, all right. One thing I could do (their method in the white garden at Sissinghurst Castle) is to plant it out from a pot annually, each spring (from cuttings taken the previous June), treating it as a climbing bedding plant. I can keep my plants through the winter in the garden, but they make too slow a start thereafter. Doubtless it would help if my soil were light.

Cestrum parqui can be grown in an open bed in the south. It will be cut to the ground in most winters, which will mean a late start, though very prolific flowering from August on. Better as a wall shrub, perhaps. Like the last, it belongs to the *Solanaceae* and carries panicles of lime green tubular flowers from early July (if the old wood survives) till late autumn. After about 10 in the evening they exhale a heavy and exotic scent but this is exceedingly elusive. By pushing your nose into the flowers you get nothing better than a sour little smell. It will grow to 6 or 8 ft without trouble and a large specimen in full flower is a fine sight, though some will disagree. 'The flowers are not very pretty,' is Bean's comment.

It would be unfair not to include a few clematis in the non-stop category. Both *Clematis tibetana* subsp. *vernayi* and *C. tangutica*, unless you've pruned them very hard, start flowering in June, and so do the hybrids between them such as 'Bill MacKenzie'. And they go on well into October, joined by the fluffy seed-heads of the early blooms. These all carry yellow lanterns. *C. viticella* subsp. *campaniflora* has bells rather than lanterns, and this starts perhaps a month later but is very persistent; white with a faint tinge of blue. It is an endearing species. *C.* × *jouiniana* has much the same colouring but its flowers, borne in dense axillary clusters, are cruciform on a vigorous trailing sub-shrub. It doesn't climb unless helped, but is ideal tree stump cover. Essential, here, to procure the early flowering clone 'Praecox', which gets going in late July and then flowers on for nearly three months.

The tenderest hebes are those with the longest flowering season. 'Midsummer Beauty' is reasonably hardy and grows into a vast, loose shrub, 6 ft high by double the width, but its scented lavender spikes come in flushes, like a repeat-flowering rose, rather than giving a non-stop performance. July and then September–October see the main displays, but worthwhile blossom will often be carried long after that.

Large-leaved hebes with a good deal of *Hebe speciosa* in their blood have the most continuous flowering season as well as the boldest and most dramatically coloured flower spikes. They are well worth the slight effort required to ensure that stocks are not lost in winter. Cuttings root so easily, at any season, even in a glass of water on a windowsill.

Fuchsias are a subject in themselves. Which of them you can grow as free-flowering, hardy plants in your garden will depend on its climate. Some will survive from year to year but will come into flower too late to be useful. Of the larger-flowered kinds I have found 'Lena' to be just about the best value of the lot. Its calyx is blush white, the double corolla purple, and this is a bold juxtaposition. The plant grows only 2 ft or so tall and is of a slightly arching habit so that it leans pensively over paving. With me this is in flower by early July and there's never a pause till the frosts arrive. 'Eva Boerg' strikes me as being slightly handsomer, and as far as I know it has all the qualities of 'Lena'. It just so happens that 'Lena' is the one I've always grown.

Fuchsias blend with abelias very happily. I have the blush white *Abelia × grandiflora* next to the vigorous red and purple *Fuchsia* 'Mrs Popple' and they both grow to about 3½ ft, the latter from ground level each season. I thin out only old, weakened branches on the abelia and it carries its little trumpet flowers from July onwards. Hardiness has never been a problem but it is the principal anxiety with abelias as a group. *A. schumanii*, for instance, would be a thoroughly desirable garden plant on grounds of showiness, for its flowers are a definite shade of pink and are an inch or more long (a mere ¾ in. in *A. × grandiflora*), but it is over-liable to be cut back by winter frosts.

However, there is a hybrid between this and *A. × grandiflora* called 'Edward Goucher' which is distinctly promising. Its flowers are a good strong mauve pink and the plant appears to be hardy, at any rate at Hilliers' nurseries in Hampshire. It was raised in the States as long ago as 1911, but has only lately reached us here.

Mention must also be made of *A. chinensis*, which is one of the parents of *A. × grandiflora* and almost as hardy. Of comparable vigour, too. Its flowers are white and quite tiny but borne in large panicles and they have the major asset of being sweetly scented.

The shrubby potentillas give tremendous value and they are so hardy. I can't think why one doesn't see them used more for hedging, as they will actually benefit from an annual clipping and some will reach a height of 4 ft comfortably. 'Jackman's Variety' and 'Katharine Dykes' would both be suitable yellow-flowered cultivars. Unpruned, many will be out by late May, but they may then spend their energies before the end of the season.

Vigorously growing young plants that have old spent branches removed to keep them active will give best value. Heavily pruned plants will not start flowering till July.

These potentillas are mainly derived from *Pontentilla fruticosa* and closely allied species. They have single flowers like small wild roses, usually yellow or white, though pink, orange and red shades are creeping in now. Such as 'Red Ace', 'Sunset' and 'Tangerine' give excellent value in a chilly, sunless English summer, or in any year in Wales, Scotland and Ireland, but as soon as the weather cheers up they turn pasty. Don't overdo the shrubby potentillas. They are good servants but a trifle boring.

There's no need ever to be bored by hydrangeas, but then they have a great range of flower types. Two of them I find outstanding for their persistence. The white hortensia 'Mme Emile Mouillère' has its first main flush in July, these flower heads being developed from the previous year's old wood and overwintered buds. But the secret of its great success is that it quickly develops and brings to maturity new flower corymbs borne terminally on strong young shoots of the current season. In this way, given a frost-free autumn, it can still be in full bloom in mid-November.

'Mme Mouillère' does not always die gracefully – her heads may get scorched brown or they may simply turn a rather unattractive pink with age – but dying is the great *forte* in *Hydrangea* 'Preziosa'. Its small bun heads start colouring pale pink in July and they gradually intensify to deep crimson, this colour being retained for many weeks. Meantime quite a succession of later developing young heads is produced, though in nothing like the quantity of 'Mme Mouillère'. 'Preziosa' gives of its best in full sun. It grows no taller than 3 ft with me, but then I prune out its old, weak, lanky shoots on a regular annual basis.

It took me some time to recognize *Magnolia grandiflora* when I first saw it growing in north Italy as a wide-spreading pyramidal tree of impressive proportions. But this was in August and there wasn't a blossom left. All had run to seed. Our climate in this respect has something to be said for it. As a free-standing specimen *M. grandiflora* never attains any great size in Britain, but however we grow it we can expect a succession of its great cream chalices, with their marvellous lemon scent, from early July till the frosts, sometimes well into November. Most of the best cultivars and hybrids such as 'Ferruginea'. 'Goliath' and 'Maryland' are precocious flowerers if propagated vegetatively. From seed you may have to wait a lifetime for the first bloom.

Finally a word must be put in for winter flowers. Most of these have a long season because the cold weather holds them back. Chinese witch hazels cannot be expected to give much more than a month or six weeks in bloom because they have just the one flush of buds opening all together. The wintersweet, *Chimonanthus praecox*, develops its blossoms over a longer period of ten weeks or so, but such as *Iris unguicularis* (*I. stylosa*), if you have a strain whose season starts in the autumn, will give you buds to pick for the house from early November till late March, whenever the weather turns mild. Of the shrubs, *Viburnum farreri* and *V. × bodnantense* spread it out and *Mahonia japonica* is outstanding. Its first flowers may be sniffed in October; they'll reach a peak in the new year and still be making some contribution when they've already been practically forgotten in the tide of spring flowers.

These winter stalwarts scarcely need the special recommendation of a prolonged season since we all grow them anyway, so highly do we value any flower that can be enjoyed and picked in the year's most forbidding months.

PROGRAMME FOR THE GREYS

Grey foliage plants are fashionable. They are generally allowed, even by those who do not relish them, to be in good taste. Fortunately they are a great deal more than that. Their pale, ghostly colouring has a luminous, light-collecting quality which draws the eye towards them. There dwelling, it discovers an endearing softness of texture (grey leaves being always woolly or silky, by virtue of the hairs that clothe them). It notes a sheen, a sparkle. Sometimes, as in *Convolvulus cneorum*, this is intrinsic; at others superimposed by dewdrops which naturally crowd upon rough surfaces and which are released only reluctantly and hours after the morning sun has played on them.

There being no great variation in colour within a grey-leaved plant, we are left free to note its other attributes: its texture, its leaf shapes, often lacy; its leaf movements and light-catching properties, notably in poplars and willows, the willow-leaved pear, the oleaster, *Elaeagnus angustifolia*; and the undersurface of such mobile leaves as clothe the long-stalked silver maple, *Acer saccharinum*, and the weeping silver lime, *Tilia* 'Petiolaris', with its prototype *T. tomentosa*.

The deciduous shrubs and trees of this genre are at their most silvery in spring, but our greatest attraction to grey is born of the heat and glare of summer, to which they offer such a reassuring contrast. But also in autumn. I have never been quite able to explain why it is that evergrey border plants in autumn are so particularly satisfying. No longer the arbiters of acquiescent refinement, as in summer, they now, in the mellow glow of a light so different from their own, take on an added sharpness, a piquancy that suggests an altogether stronger personality than we should have thought them capable of during the dog days. Partly this is because plants like *Santolina pinnata* subsp. *neapolitana*, *Centaurea gymnocarpa*, *Ballota pseudodictamnus* (refurbished with new shoots after a timely dead-heading, let's hope), *Senecio viravira* and *S. cineraria* have attained by October their climax of mature yet ebullient growth.

They make a splendid setting and context for other colours. The fulminating red of *Lobelia cardinalis*, the bright, insistent pink of *Nerine bowdenii*, the softer mauves and rosy mauves of Michaelmas daisies. And one of the most effective partnerships I have ever stumbled upon is of the single white Japanese anemone against a background of a voluminous *Artemisia arborescens*. White against grey when applied with obvious intention smacks of self-consciousness, and it is overdone in the numerous white and grey borders and gardens that have proliferated of recent years. A good idea is flogged to death. You long for sparks of relieving colour. I visited the grey, white and green garden at Sissinghurst Castle one autumn when its plant of the aromatic *Rosa serafinoi* (planted by a seat) was covered in tiny scarlet hips. In another setting they might have passed unnoticed; here, they made the scene for me.

Some greys reach a peak in autumn but all go into a decline in winter, when grey skies exploit their gloomiest aspect. In spring they have to make a new start.

I now want to discuss the practicalities of growing grey-leaved plants so that we can understand how to coax from them their full potential. Rather a case of do as I say rather than do as I do, I'm afraid, as I garden on heavy, moisture-retaining clay, which is far from ideal but, lavender apart, I don't fare too disgracefully.

An open, arid site on stony, ill-nourished ground, blazed upon by sun, eternally swept by wind – not arctic winds but, in this country, the south-westerly variety – that, in a nutshell, is the ideal recipe for all those grey-leaved plants, mostly evergreen shrubs, that come from hotter countries than our own. Their leaves, being covered with hairs, are protected from sun and wind and from undue water losses.

If you haven't a tree or a hedge or wall in your garden (and preferably no house) but are nevertheless protected by the lie of the land and distant shelter from the north and the east; if, furthermore, you have a desperately free-draining sand or chalk soil, then you have a wonderful opportunity for growing nothing but baskers, and greys will comprise a large proportion of these. They won't necessarily just be foliage plants, either. Plenty of them will gladden you with a tremendous burst of blossom: cistuses (*Cistus* × *skanbergii* is grey and *C. albidus*, which takes its name from its foliage) and helianthemums, sages purple and Jerusalem sages yellow, *Convolvulus cneorum* with white flowers, *C. althaeoides* subsp. *tenuissimus* with pink; pinks themselves in pink and white and their cousin *Lychnis coronaria* in brilliant magenta. Gazanias for months on end will unfurl and expand in a galaxy of colours every time the sun beckons.

All these and so many more invite a kind of gardening that I should love to practise, wherein I should allow no kill-joy trees, except perhaps a few small ones on my north boundary (read south for north if you live in Australia), where they could cast no shade. And I should want an oleaster, shimmering with the heat and imparting gusts of sweet fragrance on the air from hosts of tiny blossoms each June. Nothing taller than a 4-ft cistus or phlomis would gain entrance. The garden would undulate and billow between this height and nothing at all, with flat pale flagstones or expanses of shingle emphasizing the glorious, life-giving heat and glare. If yours is a fair skin and you turn to lobster red, peeling incontinently in the sun, that's no use. For you we'll have to plan a cool, dripping, fern-clad grotto – another time, not now.

It's no good asking me, as so many do: 'Which are the best grey-leaved plants for growing under trees?' because you're simply working against nature. Either you must cut down the trees or you must give up the idea of greys. Some of them will exist under trees, but that's about all. Scrawny, drawn and far greener than grey, they'll eye you reproachfully with unremitting bile, and serve you right.

Another mistake, easily made by good-natured, kindly people, is to feed your greys, offering them nourishing dishes of garden compost, farmyard manure, sewage sludge and suchlike delicacies. They are not cut out for this kind of rich living. Again their leaves will turn green, they'll grow much too fast making unhealthy, watery shoots and at the first touch of cold weather they'll die.

It's almost immoral that starvation should be what they actually thrive on, and I can imagine that a specialist in greys might become a rather callous, case-hardened type of person in whom warmth and human sympathy had withered. He would expect any house guest to exist on dried peas and soya beans washed down with Arak.

Hardiness is always a worry with the greys – hardiness in the face of damp and cold instead of the dry and only moderately cold climates they are cut out for in such native areas as the Mediterranean and South Africa. The harder you grow them – by which I mean the poorer the soil, the more open the site – the better will they resist and survive any cold that's going around. If you can't be certain of adequate drainage at all seasons, you can get round it by raising your beds a foot or so, and retain them in a simple and dignified manner with courses of not visibly cemented bricks or stones.

Some greys, like the South African *Helichrysum petiolare*, are nowhere near approaching sufficient hardiness in our climate to be trusted to come through a winter outside, even though they may succeed in doing so on occasion. But they are easily propagated and not too expensively overwintered as rooted cuttings under frost-free glass. Furthermore, once planted out or in the tubs, ornamental pots or window boxes that suit their flowing lines so well, they grow at tremendous speed. In a case like this, where the container ensures a restricted root run and hence a limited food supply, I do recommend using a strong compost like John Innes No. 3, and watering them regularly and thoroughly. Otherwise you'll find their leaves scorch and they'll get a tired look before the summer's half-way through. Your waterings will leach out the nutrients so you should apply a liquid feed every week from early July onwards.

As I have already hinted, deciduous trees and shrubs with grey leaves or with grey undersides to their leaves are quite uncharacteristic of the general run of greys. They still need full exposure but they are hardy and their soil requirements are not specific. Willows, in fact, like it rather wet. There are two native willows well suited to our purpose and they are specially valuable to gardeners in cold localities. *Salix lanata*, the woolly willow, is easier to please in Scotland and the north than down my way. As seen in cultivation it has a broad, rounded leaf and it is a shrub that you expect to grow no more than 3 or 4 ft high and rather more across. Male clones have attractive pussies in spring. *S. alba*, the white willow, is an enormous tree but the variety *sericea* is quite manageable. Pollarding them hard each winter, I have a group of three towards the back of my Long Border. Behind them is Dickson's golden elm, which I also prune not to grow more than 15 ft tall, and in front a *Rosa moyesii*, whose scarlet hips show up well in autumn against the grey; and a group of tall blue veronica. I enjoy this piece of my border unfailingly.

Some comments on the pruning of the more popular and reasonably hardy grey-leaved shrubs may be helpful. If their foliage is the main object, this will develop most tellingly if quite hard cutting back is practised regularly each spring (though there are exceptions). Such treatment often eliminates flower production, and that may be your very aim. Suppose you are growing *Santolina chamaecyparissus* (*S. incana*) or the even more lax-habited *S. pinnata* subsp. *neapolitana* as a low hedge surrounding a parterre or herb or knot garden; the more formal it looks the better and you certainly won't want it to flower. A 'short-back-and-sides' trim each spring, then, will prevent flowering and keep the outline neat. Santolinas are not out-and-out hardy and will not long survive as a complete and presentable hedge without gaps, in frosty areas. That was what they found in the parterres at Drummond Castle near Stirling. They are situated in a classic frost hollow, which is why you have such a magnificent view down on to them from the castle. They have substituted *Helichrysum splendidum*, a compact South African shrub which is astonishingly hardy. Left to itself it will grow 4 ft tall, but it takes to regular pruning extremely well and the small rosette formation of its young shoots is most attractive.

I should at this point perhaps warn that after cutting a grey-leaved shrub hard back, the young leaves on lush, sappy shoots that next appear will be green, not grey. This is nothing to worry about; as they mature they soon grey up. Another nasty situation develops when you're showing a visitor round the garden during or just after rain. Again the greys turn green or grey spotted green. The garden I'm writing of in this chapter is a sun garden for sun worshippers, plants and humans. You have no business to set foot in it except under bright or breezy conditions (I am not writing of my own garden, where visitors have to tramp round in a monsoon, if need be).

That is, unless you have to prune your Jerusalem sage, *Phlomis fruticosa*. This is notorious for giving everyone, even though not generally that way inclined, an extremely unpleasant dose of hay fever. Not only do your eyes stream but it grips you in the throat so that you cough and choke and can scarcely breathe. This shrub needs a fairly hefty trim over after flowering, going back on each shoot behind the flower trusses so that you remove a bunch of them including a healthy, terminal leafy shoot, all at one go. Do it when the plant is wet and all will be well.

A comparable shrub for size and performance is *Brachyglottis* (syn. *Senecio*) 'Sunshine', a hybrid erstwhile confusingly lurching between *B. laxifolia* and *B. greyi*. I like this to flower, so I cut out its flowered branches in July, in the same way as the phlomis, leaving its unflowered sprays intact. They will give me blossom the next year. Those who resent its flowering, however, regularly trim it over each spring. That other popular grey senecio, *S. cineraria* (syn. *Cineraria maritima*), grown for its foliage, really does carry too miserable a ragwort flower to be tolerated. Cut this into bare wood each April and you'll not be troubled.

But whereas you can treat most greys in this manner and get away with it, a few are tricky and reluctant to respond to the knife of the master. *Senecio viravira* cannot be depended upon to overwinter anyway, but if it does, its response to cutting back is patchy at best, sometimes lethal. *Centaurea gymnocarpa*, my favourite among all the greys for its long, doubly toothed leaf, may overwinter successfully twice out of three times but then its main object in life becomes the production of its small thistly purple flower heads. Rather than cut the shrub back, which kills it, inexorably, wait till the flower buds are quite forward and about to disgrace you. Then follow them back till you find where you can cut out large clusters at one go. Having treated the bush so that none remain, it will still have some foliage shoots on it. Subsequently, it will (if you have a satisfactory strain not raised from seed but kept going from cuttings) make new foliar growth and give up all attempt at flowering for the remainder of the summer and autumn.

Now for propagation. The main problem, here, is to prevent rotting. Any soft, leafy young shoot covered with a thick felt of hairs has a way of collecting and holding water. All the more so under conventional methods of propagation which favour a humid environment so as to preclude water losses and wilting. Grey-leaved plants do not propagate well under mist unless fungicides are constantly included in the spray. Neither should they be given close conditions except for a very short initial period. Allow some ventilation to the frame or propagating unit in which they are set, at least after the first week. Not so much that the cuttings wilt badly and cannot plump up again at night. You'll soon learn to get the balance right.

In selecting your material for propagation, avoid soft shoots from rapidly growing young plants. For example, if you take the young shoots from *Senecio cineraria* in late May following a hard pruning in early April, they'll be much too lishy. Wait till August or September and they'll be just right.

It helps to immerse your cuttings in a fungicide solution before inserting them. Subsequently we always spray once a week with a range of fungicides as a matter of routine, so that any moulds hovering around cannot spread. Varying the fungicide keeps different strains of fungus on the hop. You must also inspect your cuttings every few days for the first three weeks or so, and promptly remove any dead leaves. Try not to let cuttings go to bed in the evening with wet leaves. Do your watering in the morning. Once rooting has taken place (and the use of a rooting hormone hastens this), the cuttings can be allowed full ventilation and your worries are at an end.

When you plant your greys out in spring, which is the normal time for establishing them, you are likely to run into trouble from sparrows, which pick off all the leaves and take them to their nests. Black cotton works infallibly against this pest.

One last word. Don't cling on to old plants that have lost much of their vitality. A young, vigorous lavender, for instance, will flower non-stop for three months instead of the usual three weeks. *Artemisia arborescens* is another short-lived grey that goes into a protracted decline without actually dying. Replace it every third year.

SORTING THROUGH ERYNGIUMS

A surprisingly large number of visitors to my garden ask me, each summer, to identify eryngiums for them. A blue thistle is how they describe it. Any prickly plant is likely to be so termed. But eryngiums are among those hoodwinking genera that belong to the *Umbelliferae* (now *Apiaceae*). Astrantias and aciphyllas are others. Most umbellifers proclaim their identity at a glance. Parsley, parsnip, celery, fennel, carrot, hemlock, hogweed – poisonous or edible, they all have the same unmistakable kinship. Eryngiums all look like eryngiums and that's the most you can say of them.

They are sometimes unhelpfully lumped together as sea hollies. Only our native *Eryngium maritimum* inhabits the sea dunes and shingles, while the holly is an even more distant relation than the thistle.

Eryngium × oliverianum is the plant that principally catches the public's eye in my borders. It has a brilliant, metallic blue sheen on its stems as well as bold flower heads. Each of these consists of a domed centre packed with tiny blue flowers and framed by a ruff of stiff and painfully spiky bracts, all blue. Each principal head measures 4 or 5 in. across but is supported by a number of smaller subsidiaries. Bees, butterflies and orange soldier beetles visit them (and so do ants), but I can detect no scent. The plants grow 3½ ft tall but although their stems are rigid they sway over from the base. Being prickly in all their parts, the task of providing discreet support with short pieces of cane, and string looped from stem to stem, is generally accompanied by a ground swell of imprecations.

The Old World eryngiums of which this is typical all make a basal tuft of foliage in one year on which they flower in the next, but not again. So if, on a young plant (where you notice what's going on more readily than in a complex old colony), you are enjoying a flowering stem one summer but can see no tufts of basal foliage, you may be sure there will be no flowers in the following year. Once your colony is well established there should be a reasonable spread of flowers every year, but even so, the amount of blossom can vary quite a lot.

Below ground, this group of hardy eryngiums is always tap-rooted, which means that they accommodate most happily to light, free-draining, stony soils. Indeed, if you had a seaside garden with soil of the poorest, shingly quality, this is one of the genera of which you could make a great feature. The very poverty of the soil and the openness of the situation would ensure short, strong stems for which no support would be necessary, despite the wind's violence. Furthermore, your flowers' colouring would reach its greatest potential brilliance. Eryngiums always lose in quality and colour if at all shaded.

But to return to what goes on underground: their long, thick tap-rooted system means that they take a little time to establish but seldom, if ever, need moving thereafter. On heavy soils like mine these roots are frequently killed by a smelly, slimy bacterial disease, and if this should happen to strike in summer, the flowering stem above will collapse. It is never a wipe-out disease but breaks out sporadically in one part of a colony but not another.

As so often with fleshy-rooted plants, the easiest method of propagation is from root cuttings. Two-inch lengths of root can be potted up in a cutting compost during the winter and kept in a cold frame until they are growing strongly, by June, when they can be lined out to grow on. By the autumn they'll be fit to cope with border life.

Eryngium × oliverianum is a splendid border plant which will associate with most colours, other than blue. Its own colouring needs lifting by another in contrast. I am pleased with the effect of having it next to the golden daisies of *Heliopsis*, but I can also visualize it associated with montbretias, whether yellow, orange or red. They would do the trick.

This sea holly has a July–August season. A hybrid called 'Violetta' is comparable, though shorter in the stem. Its colouring (stem, flowers and bracts) is indeed a sumptuous violet but, alas, its constitution is weak and it is particularly prone to a debilitating bacterial leaf spotting.

Eryngium alpinum flowers in June in the south and is distinct from all others. Its height is 2½–3 ft and it is moderately sturdy in the stalk. Large, domed blue heads are framed by intricate and lacy bracts in several series. And they are soft to the touch. Lulled by this novel experience, you find yourself going into a trance while passing your hand repeatedly up the flower head.

If you meet a clone called *E. alpinum* 'Improved', you're sure of a good plant (provided it has been propagated from root cuttings and not from seed) as its colouring is more definite than the unselected species.

Some eryngiums have small inflorescences, only an inch or two across but making up for size in numbers. *E. × tripartitum* (a hybrid, as is *E. × oliverianum*) is the pick of the bunch. *E. amethystinum* is likely to be the same thing. It is azure blue throughout, 2–2½ ft tall with widely branching stems. These are weak at the base and flop sideways but if the plant is grown in a really open situation, as it might be in an island bed with little or nothing taller than itself to draw it, and the soil rather poor – then it'll look very acceptable, flopping or no.

Eryngium planum has small heads more closely packed, not widely branching, and grows to 3 ft without ever heeling over. But its colouring is a darker, dimmer blue. Its shorter-stemmed offspring, 'Blauer Zwerg', insults us by being a flopper after all. There's no point in growing this, as I quickly discovered.

Two dwarf, 1-ft species are worth growing mainly for their foliage. *E. bourgatii* has crimped, fairly divided, slightly variegated leaves; *E.*

variifolium has them strongly marbled in green and white. Its bracts are outstandingly stiff and spiny. I don't get excited by either of these but they would look well if the largest among smaller-scale plants.

Then the sea holly itself, *Eryngium maritimum*. Again it is an excellent plant in an open situation (only a foot high) and very likely the chief reason for its being so seldom offered is that, being a native, it is disregarded. Its bracts are pale and silvery and make a great impression. The domed inflorescence is mutedly blue. Our only other native, *E. campestre*, is rare. In this country I have seen it only in the walled garden at Wakehurst Place. I admired it and was given a piece. I have subsequently taken a scunner at it in Hungary, where it grows in every piece of permanent grassland and is a most uncomfortable plant to sit on; at those very moments, furthermore, when you would particularly like to relax in comfort. *E. campestre* is green throughout, pretending it isn't there.

The 3-ft *E. giganteum*, often known as Miss Willmott's Ghost (I presume she was pale and prickly), differs from the others in being monocarpic. Having flowered, it dies, and you must depend on self-sown seedlings to carry on the race. This is easy once a colony has been established, but needs a little thought and concentration in the early stages. The plant is bold, stiff and handsome with blue-green flowers surrounded by magnificent stiff, silver bracts. Many plants develop virus disease, which is easily spotted before they reach the flowering condition. The leaves become much attenuated and are violently mottled. They make miserable flowering plants and should be eliminated as soon as the trouble is identified.

The specific epithet *giganteum* has deceived many a gardener, this plant being in fact a mere 3 ft tall. Wait till we reach *E. pandanifolium*.

All the sea hollies so far described are good for cutting, and especially to dry for winter arrangements. In that case, be sure to take them as soon as they have reached their prime. Don't think to enjoy them in the garden for a few weeks and *then* cut them for drying. They won't be worth it and you'll be paid out for your greed.

The New World eryngiums are a very different kettle of fish. They come to us from the general area between the S.E. States and Uruguay. All are evergreen, forming loose rosettes or units of stiffish, strap-shaped leaves with spiny margins. Indeed, their growth rather resembles that of the pineapple and other bromeliads. Their flowers are in more muted colours than the Old World brigade, but the plants themselves are structurally bold and arresting. As they come from warm, even tropical latitudes, they are always expected to be tender in this country. In fact, they are a lot hardier than a timid gardening public gives them a chance to demonstrate.

Unfortunately their naming is in a state of extreme confusion, the same name being frequently applied to several different species, which is the most maddening situation that can possibly befall the hapless gardener. He cannot be sure what he's buying unless he sees the plant.

The one I was given as *E. agavifolium* has also been marketed as *E. bromeliifolium*, is said by Kew to be *E. paniculatum*, but that is a synonym of *E. eburneum*. It is a nice plant, even so, making a large cluster of narrow-leaved crowns about a foot or 18 in. high. The inflorescence rises in July to 3 or 4 ft, and is stylish with the principal aggregate of small green flower heads radiating from spokes at the top of the stem but supported by subsidiary clusters along the stem's length. There are no conspicuous involucral bracts in this or the majority of the American species. Seed is an easy method of propagation.

The true *E. agavifolium* and *E. serra* are closely related. The former has broader leaves than most and larger thimble heads to 4 ft. It is a coarse plant which I mildly dislike. If I have seen *E. serra*, and I probably have, it hasn't registered.

With *E. pandanifolium*, however, I am very familiar, having grown and loved it in my mixed borders for twenty years. It is reputedly tender, but there were bits to carry me on even after the once-in-a-century winter of 1962–3. Given the freest drainage and protection from cold winds it will thrive even in the midlands. John Treasure, at Burford House Gardens, Shropshire, which are in a frost valley, grows it in stone paving near his house.

The plant has 6-ft long, spine-edged scimitar leaves in a dense thicket of crowns and nothing much happens till August, when the flowering stems first appear. They branch into a vast but elegant candelabrum, 8 ft high in my garden but often only 5 ft, consisting of quite tiny, ½-in. long cones of flowers, which open from late September onwards. They often remain in condition till the year's end, if frosts hold off. The flowers themselves, in my strain, are a very soft, restrained shade of mauve, but I noticed in the Chelsea Physic Garden that their plant is purple in the bud and becomes a darker purple still after flowering. They label theirs *E. pandanifolium*, and this was the name under which it received the R.H.S. Award of Merit when Mr Harold Hillier showed some giant stems in 1976.

Anyway, to continue my tale, this species, whatever its name, collects a lot of dead foliage which I usually clean out (wearing gloves to protect me) in the spring, but after ten or twelve years my group became so thick that it was really not as beautiful as of old. Heart in mouth I decided to lift, split and replant. What was my surprise on discovering that the task was easy. Unlike the Old World eryngiums, this group has no tap roots but an ordinary system that readily transplants. Neither can you propagate from root cuttings. Division or seed is your way. *E. pandanifolium* ripens its seeds only after hot summers have induced it to flower early and autumn frosts hold off Sown in the following spring, germination is abundantly free.

E. proteiflorum is a comparative newcomer from the highlands of Mexico and it seems to be hardy. I first met it flowering in the gardens at Inverewe, in north-west Scotland, so it evidently does not require an extra dose of sunshine to make it perform. Its basal leaves stay near the ground and give rise to 4-ft-tall, sparsely branching stems of great distinction. The spaces between flowering stems are wide and noticeable, but the heads themselves are large with a ruff of conspicuous white bracts (Old World sea holly style) surrounding a blue dome. This comes easily from seed. It is amusing to note that when the seed house of Thompson & Morgan introduced this species in their catalogue, they mistook the name of the authority who described *E. proteiflorum* for the name of the plant itself. Confusion worse confounded, it goes around masquerading as *Eryngium* 'Delaroux'.

THE PAMPAS AT HOME AND ABROAD

We have only to see it flowering in countless front gardens each autumn to appreciate the widespread appeal of pampas grass, *Cortaderia selloana*. A noble plant and yet too often, in cultivation, flawed. 'The garden-plant has a sadly decaying, draggled look at all times and to my mind, is often positively ugly with its dense withering masses of coarse leaves, drooping on the ground, and bundle of spikes, always of the same dead white or dirty cream-colour.' Thus W. H. Hudson, field naturalist and author, whose early life was spent in Argentina in the middle of the last century. From him we can get a vivid image of how this grass grew and looked on the great plain of La Pampa, where it comes from.

It flourished on moist clayey ground where its spears often attained a height of 8 or 9 ft. 'I have ridden through many leagues of this grass with the feathery spikes high as my head, and often higher,' Hudson tells us in *The Naturalist in La Plata*, first published in 1892. 'It would be impossible for me to give anything like an adequate idea of the exquisite loveliness, at certain times and seasons, of this queen of grasses, the chief glory of the solitary pampa.' But he succeeds in doing so.

'Colour – the various ethereal tints that give a blush to its cloud-like purity – is one of the chief beauties of this grass on its native soil; and travellers who have galloped across the pampas at a season of the year when the spikes were dead, and white as paper or parchment, have certainly missed its greatest charm. The plant is social, and in some places where scarcely any other kind exists it covers large areas with a sea of fleecy-white plumes; in late summer and in autumn, the tints are seen, varying from the most delicate rose, tender and illusive as the blush on the white under-plumage of some gulls, to purple and violaceous. At no time does it look so perfect as in the evening, before and after sunset, when the softened light imparts a mistiness to the crowding plumes, and the traveller cannot help fancying that the tints, which then seem richest, are caught from the level rays of the sun, or reflected from the coloured vapours of the after glow.'

Clearly we shall not often be able to do full justice to such a plant in our English gardens, but we must try as we may to mitigate its faults and enhance its glories.

First, how shall we place it? A plant 10 ft tall and 6 ft across is not easily accommodated in a small front garden without looking self-conscious. It is often planted, like a specimen tree, to make an isolated lawn feature. But this is to draw attention to its foliage which, although a perfectly fitting plinth to the plumes above it, is a weak feature on its own and in the non-flowering seasons. Neither does the average suburban house make a suitable background. A dark, plain background of trees or evergreen foliage, on the other hand, will throw the white panicles into relief, and so to one side rather than in the centre of your garden is likely to look right. A screen of pampas dividing garden from compost and rubbish heaps could also look good. Beware allowing it to overhang paths where legs swish by, for the margins of pampas leaves, technically described as scabrid, i.e. rough to the touch, are sharp, saw-edged and painful. You never want to handle the plant without wearing gloves.

It looks splendid overhanging and rising above the margin of any reasonably sized pond, in which its stateliness will be reflected. Obviously a landscape garden setting will suit it best of all, but it is surprising how seldom you see full justice done to it even there. Instead of large plantings that can be admired as bold features from a distance, you see solitary plants dotted about as though the gardener's mind were incapable of matching the plant's high style.

Then there is the colour question. Hudson's emphasis on this seems to arise as much from the quality of the light falling on the plant as from its own intrinsic colouring, although the latter could be expected to vary considerably as between thousands of plants, each of which was a seedling having a different pattern of genes from its neighbours. And different plants could also be expected to be at slightly different stages in their flowering. This again would lead to colour variations.

As regards light quality, we can help matters along by siting our pampas plants where they are struck by early morning or by evening sunlight. But it would be unwise to make a mixed planting of seedlings. You only have to look around you to appreciate how very inferior large numbers of these are to some of the selected clones that have been named and are propagated vegetatively, although, as an aside, I should warn that the wrong plant frequently masquerades under the desired name. Probably the most sensible course is to mark down a handsome pampas seen at flowering time in a garden or on a nursery, and to beg or buy a piece from that in the following spring, which is the season when sensitive grasses can be disturbed without their taking undue umbrage.

I have never thought highly of the pink- and purple-tinted pampas that go under names such as 'Rosea' and 'Violacea'. They may look attractive at close quarters but are lost and wasted at a distance. Incidentally, the sexes in *Cortaderia selloana* are on separate plants and it is said that the females are much the handsomer. Not being practised in pampas sexing I have to take this assertion on trust.

Of the named clones, 'Sunningdale Silver' is one of the most striking and it received a First Class Certificate in 1971, following a trial of perennial ornamental grasses at Wisley. It grows 10 ft high and the inflorescence is 18 in. across by 2½ ft long, loose, plumed and silvery white. This pampas is grown by the lakesides in Sheffield Park gardens in Sussex and is at its best in October, when it strikingly contrasts with the autumn colouring for which those gardens are famed above all others in Britain. But the dark backgrounds of conifers and rhododendrons are also vital to the effect. I could only wish that the pampas were planted there in far greater numbers. Another variety grown at Sheffield Park is the imposing, though strangely named 'Monstrosa'. There is nothing even marginally monstrous about it, but its plumes are terrific, though light and open-textured.

The only pampas I am growing (so far) is the comparatively dwarf 'Pumila', which suits a smaller garden scale. It grows 6 ft tall and its spikes are all upright in a dense but tapered brush, very numerous and closely set and hence easily picked without gaps being noticed.

I have never seen pampas in their popular role as cut decoration for winter (and thereafter permanent) use that didn't look pretty horrible, but if you must join in this game do at least pick them when so fresh that they have not yet fully expanded. Thus they won't have become tarnished, bleached or battered through exposure to the weather. Most people (not greedy but liking a lot) prefer to enjoy their pampas in the garden for a few weeks before bringing them in. By then they are too late.

And in the garden itself the panicles are usually left to become unhappy, derelict grey ghosts, gradually disintegrating and not finally tidied away until the following spring. Then (and the pundits encourage this) a match is put to the plant, the owner stands back with childlike glee and the whole mess goes up in flames, leaving a different but equally horrible mess behind; a charred eyesore for weeks until new growth conceals it. To cut the whole plant over neatly is far better, and only a slightly greater effort. I have of recent years taken to doing this in early winter so that its flowering remnants shall not linger on.

There are cases where pampas grass, when grown in Scotland and the north of England, fails to develop its flowers before the onset of winter; others where it succeeds. I'm sorry not to be able to give details and recommendations of clones for northern gardeners. However, there is a most satisfactory cortaderia that is much seen in the north: the toetoe from New Zealand, *C. richardii*. It flowers early and regularly in July and August. Its flower heads are not as long as they usually are in the pampas and they have a tinge of yellow in them. Eight or 9 ft tall is usual and they lean outwards obliquely from the centre of the plant. Handsome, but more bedraggled than most after a battering by rain and wind, so do tidy them over before this has gone too far.

Pampas present no difficulties in cultivation provided they receive plenty of sun. They are often, nowadays, produced for sale in plastic polybags and can then be planted at any season. But if their roots have to be disturbed, the dormant season – autumn or winter – is fatal. Spring's the time to split clumps up and even then it is important not to let their exposed roots dry, and the replanted pieces must be kept well watered until clearly re-established. After that they'll be as drought-tolerant as a cactus.

FLOWER OF THE SENTIMENTALISTS

Blue is the flower colouring that most surely gets us, emotionally, and the forget-me-not touches the flower lover's heart in its most nostalgic and associative spot. 'This pretty plant is peculiarly the favourite of poets and sentimentalists,' Sowerby debunkingly remarks in the 1867 edition of his *British Flora*, but it is their favourite because it is ours and we are all sentimentalists even if we cannot rise to poetizing.

The name forget-me-not has been attached to the genus *Myosotis* since the fifteenth century, it seems, which is long enough for most of us, but there is an uncomfortable theory that it may previously have been applied to a species of bugle (*Ajuga*) with foliage of a nauseating, not-to-be-forgotten flavour. People were very childlike in those days: liable to put anything and everything into their mouths.

Forget-me-not was a translation of the Old French *'ne n'oubliez mye'*, but the plant is nowadays called myosotis in France, and very often in England, too. You would never hear a parks superintendent referring to a forget-me-not, now would you? Myosotis every time. This word is derived from two Greek words meaning a mouse and an ear. The myosotis leaf is supposed to resemble a mouse-ear. Even allowing for the fact that there are many species of *myosotis*, and many species of mouse, the resemblance is unimpressive and we are not surprised to learn that myosotis was a name originally applied to some other, unspecified, plant.

And why forget-me-not? Because the wearing of the flower ensured that its wearers would never be forgotten by their lovers. But this reason is also hard to wear, quite apart from the fact that a plucked forget-me-not would start wilting within five minutes and be a sorry sort of reminder of or to the loved one. What surely stands out a mile is that it is the flower itself that is not-to-be-forgotten.

The myosotis to which this name is particularly ascribed is *M. scorpioides* (syn. *M. palustris*), the water forget-me-not (*Sumpf-Vergissmeinnicht* in German; what a splendid word). And this seems absolutely fitting, because the wide-eyed innocence is devastating in its appeal in that particular species. There is the setting: rank green herbage and the smell of water mint at the stream-side. There are the pale green leaves, and the large, pale blue flower itself, blue with a yellow eye. The poets have all been well aware that a brook should be the setting for forget-me-nots.

One of Hugo Wolf's most touching *Lieder* is to the text of Mörike's lyric, *Heimweh* (Homesickness), which depicts a traveller trudging miserably further and further from his home while he notes the changing scene. A murmuring brook tries to comfort him: 'Poor boy, come to my brink; here you shall still see forget-me-nots.' 'Yes, they are beautiful wherever they may grow, but not as there. On, on! My eyes are brimming over.'

> *(Das Bächlein murmelt wohl und spricht:*
> *'Armer Knabe, komm bei mir vorüber,*
> *siehst auch hier Vergissmeinnicht!'*
> *–Ja, die sind schön an jedem Ort,*
> *aber nicht wie dort.*
> *Fort, nur fort!*
> *Die Augen geh'n mir über.)*

But the water forget-me-not also features in one of Wilhelm Busch's strip cartoons, published in Munich in the middle of the last century. The haughty young lady, Adelen, wearing a voluminous pink crinoline, starts her ill-fated walk by stooping at the brookside to pick our flower. A frog jumps into the water; she swoons ... but there's worse to follow and her crinoline cage is finally adapted by a stork for its treetop nest.

M. palustris is a true perennial of stoloniferous habit. Wherever its stems touch down on earth or water they send out roots. Its extended season is from July into autumn, and I have never seen it more prettily displayed than when wholly intertwined at the waterside with the lesser spearwort, *Ranunculus flammula*, whose tiny buttercups are borne in clouds. That happened to be in Sheffield Park gardens but I rather think the partnership was unpremeditated.

There is a dwarf and more compact though still spreading cv. of *M. scorpioides* called 'Mermaid', and this makes an excellent bedding plant anywhere that the soil remains reasonably moist. Still on the blue and yellow tack, I have enjoyed it in the company of the self-sowing annual *Calceolaria mexicana*, which has bright acid yellow pouches on light sprays – not at all in the clumsy image of the florists' pride and joy.

A dramatic blue-and-yellow association may be seen by anyone with reasonable legs on the chaotic rocky slopes of Ben Lawers, in Perthshire, where *Myosotis alpestris* grows on ledges with the spurge-like *Rhodiola rosea* (syn. *Sedum roseum*); it superficially resembles *Euphorbia myrsinites*. To these are here and there added the carmine cushions of *Silene acaulis*, which flowers so much more freely in the wild than when tamed, that it should exclusively be left to its natural habitat.

The forget-me-not of spring bedding is a mixed-up kid deriving a good deal from *M. alpestris*. It can live on beyond its flowering season, but is seldom allowed to. With good reason, because old plants become horribly mildewed later on. There are two ways of treating these forget-me-nots in the garden. The first is to do them properly, as spring bedding to make a background for tulips. A cliché this may be, but it works and one of the nice things about myosotis (as we must call them in this context) is that they become increasingly informal as the season advances and they keep on opening new flowers along their ever-lengthening stems.

Myosotis have an unusually long flowering season, when compared with other spring contemporaries. When bedded, they far outlast the tulips. The second, cottage-garden, method of treating them, is to let them self-sow in various parts of the garden, among other plants. They can then be allowed to run their full season, undisturbed. All the gardener needs to do is to thin their inevitably overcrowded seedlings, in the course of the summer, so that those remaining have the chance to make decent-sized plants. There are few spring flowers they don't associate well with, provided these are not themselves blue.

In my own garden, I particularly enjoy them with the apricot-orange flowers of *Geum* × *heldreichii* 'Superbum'. Again, elsewhere, with *Deutzia* × *rosea* 'Carminea', whose pink wands open in the second half of May and bend down to ground level at their tips. A third association that works out is with the brilliant yellow, early-flowering broom, *Cytisus* × *praecox* 'Goldspeer'. The white daisies of *Olearia* × *scilloniensis* have made up a happy trio with these two in those years when this daisy bush has brought its dormant buds through the winter unscathed.

Whatever you do, avoid getting forget-me-nots near to other blue flowers, or there'll be an 'I'm bluer than you are' fracas that will end in tears for one party or the other. I had the low-growing *Ceanothus divergens* near the last-mentioned grouping. It is a good shrub, only 2 ft tall with evergreen foliage that survives the winter unprotected, and its flowers would have looked quite an acceptable blue had not the forget-me-nots been hard by. In the end I took the easier course and rid myself of the ceanothus.

VIOLETS

One of the gardening extras I should love to indulge in, had I but the time, is the growing of violets in a cold frame so as to have them for picking all through the winter. I can just imagine the gust of violet scent that would greet me each time I opened the frame with a view to … well, principally with a view to receiving that gust of scent as I opened the frame. And if you ask how I know the violet scent would be so generous to me the answer is that the whole plant, not just the flowers, smells of violets. And then, of course, I should pick the blooms, put down some more slug killer and perform any other little necessary ministrations, all *con amore*.

Each April, at the end of the violet season, you see advertisements in gardening journals for surplus stock from commercial growers, usually somewhere down in the south-west. That would be the way to make a start. The plants can be lined out, a foot apart and, for preference, in a cool, moist spot for the summer; then lifted and planted in your frame in early autumn. The frame, for this winter role, needs to be in as sunny a position as possible and the soil should be rich in organic matter like leaf mould or garden compost.

Although the sun will coax their blooms out, violets loathe stuffy heat, so they need generous ventilation and plenty of water throughout. Then, in April, the frame is emptied, the plants divided and the strongest crowns lined out in their summer quarters again.

You can also raise stock from seed, but this needs an extended period of low temperatures after sowing to make it germinate. Sow in a pot or box in autumn, therefore, and overwinter the container where frost can reach it. Then the seedlings can be pricked out and finally lined out the next spring and will have made strong plants by the following autumn.

There is, alas, a most debilitating pest of the sweet violet (not of other kinds) which is rife in my garden and seems to be virtually ineradicable, though one could control it by spraying under the controlled conditions of cold frame cultivation. But when you have violets that have self-sown in every part of the garden, under hedges, in the heart of shrubs (for they tolerate deep shade), in rough grass and in paving cracks, there's really nothing you can do for them and they will always remain a focus for fresh infestations. The pest that is in question is the gall midge, *Dasineura affinis*. It reduces most or even all of the violet's foliage to puffy, bloated lumps and very much weakens the plant. We are told to spray or dust with insecticide before peak egg-laying periods (the midge larvae feed within the gall and are inaccessible by then), to wit late April, early July, mid-August and mid-October 'as necessary'. How do we decide on this necessity? By the time we notice the galls we're already too late to cope with that generation. It sounds a depressingly uphill struggle.

Nevertheless, despite the midge's depredations I still have violets all over my garden and many of them succeed in flowering well. This is our native *Viola odorata*, whereas the cultivated kinds are of mixed parentage, for there are a number of species that have long been grown for their scent in Turkey, Persia and other Middle Eastern countries and they have contributed to the florist's flower.

But our own sweet-scented violet is a most versatile and variable plant. It comes in a wide range of colours, violet, purple, pink, creamy yellow, white and many intermediate shades. Some, like the pink 'Coeur d'Alsace', have been given clonal names, but I find that my pink ones self-sow as freely as any, and all the progeny are slightly different – not unnaturally, as there are violet-coloured ones close at hand. Those that turn out muddy I remove when they're in bloom. My point is that there's unlikely to be any fixed clone warranting the 'Coeur d'Alsace' tag. The albinos, however, apparently retain their pure whiteness from seed even when there are other colours around, but they may have colour potential in their genes, for aught I know.

There is a tendency for the whites to have stalks that are a fraction longer than the coloured kinds (much depends on how open the site or how drawn by shading), and for this reason and because they go on flowering a bit later than the others I seem to find myself picking little bunches of them most of all. Beware picking wet violet leaves. You always want some to frame your bunch, but if wetted all over their surface these leaves are archsiphoners of water from the vase you put them in on to the furniture it's stood on.

There are some sweet violet strains with a strong tendency to flower in the autumn and thereafter whenever the weather is mild through to March. A cosy corner helps them greatly in this respect, and violets are as happy to be baked in the open as they are amenable to deep shade. The white and apricot violets are never precocious, in my experience, but the pink and violet colours often.

We are inclined to call any mauve or bluish violet that is odourless a dog violet, but violets are botanically a complex lot and, in fact, the 'dog' violet that we most commonly find in woods in spring, and sometimes in autumn also, is *Viola riviniana*, not *V. canina*, and we should call it the common violet. Granted that it disappoints by not smelling; once we have accepted that fact we can appreciate that it does make an exceedingly pretty, showy plant and it is well worth having in the garden. I do not think we ever consciously introduced it into ours, but there it is all over the place, and especially welcome in the risers of steps, in the cracks of paving on a terrace where we sit, and peeping out from the bottom of yew hedges and topiary specimens. It seems to enjoy a good share of sun.

V. riviniana Purpurea Group has flowers of a deeper shade, and its foliage is so purple, at any rate in the strain we see in gardens though probably not in its natural habitat, as to be almost black. It self-sows and the seedlings all retain this fascinating depth of colour. Dark colours need setting off so that their darkness is not wasted in gloom, and I have not yet done much about *V. r.* Purpurea Group in my garden. Graham Thomas says it makes a good background for *Astrantia major* 'Sunningdale Variegated', whose variegation is at its brightest on the young spring foliage. Beth Chatto suggests growing the violet with Bowles's golden grass and with snowdrops, which would presumably flower before the violets (unless it was the late-flowering *Galanthus platyphyllus*), using their foliage as a background carpet.

Another thought occurs to me, that of associating *V. riviniana* Purpurea Group with a yellow-flowered violet. The one I have is *V. pubescens* var. *eriocarpa* (syn. *V. pensylvanica*), which is a very bright yellow and makes a gay contribution to the spring scene, but perhaps the European *V. biflora*, of similar colouring, would be as good and easier to procure.

The boldest and showiest of all violets is the white *V. cucullata* 'Alba'.

From Labrador and eastern Canada, it is ultra-hardy but completely deciduous; the only intimation of its continuing existence in winter is a mat of knobbly rhizomes at the soil's surface. It does not return to life till April, and its large white blooms, with mauve pencillings on the lower lip, make their display before the leaves have expanded enough to hide them. The mauve-flowered form is sometimes apt to hide itself in foliage but it can be effective. I should give it full sun.

V. cucullata seeds itself like crazy. The seedlings grow very fast and themselves have seedlings before they have flowered. Or so it appears, but the miracle has an explanation. The love life of all violets is most peculiar in that they first, in spring, carry normal flowers such as we grow them for but later, quite unobserved and hiding beneath their foliage, produce flowers without petals which, nevertheless, are fertile and seed freely. The seedlings of *V. cucullata* are too small to flower normally in their first spring but are large enough by the summer to produce a generation of hidden (cleistogamous) flowers, and these seed in the autumn.

The almost hardy Australian violet, *V. hederacea*, flowers in summer and autumn over a months-long season, and it is a dear thing which many more gardeners would grow if they realized how easily it can be treated as a bedding plant, even though stock cannot be altogether trusted to winter safely outside (though it seems to be hardy enough with me). All you need to do is lift a few pieces in the autumn and overwinter them in a frame. Planted out in the spring, they rapidly spread by overground runners and soon make a delightful flowering patch. I saw them used in this way in the Edinburgh Botanics. The flowers are pale at the margin, blue in the centre, and the lowest petal recedes, giving the impression of a chinless wonder. Again it is a deciduous violet and makes a mat of rhizomes.

BULBS TO PLANT GREEN

There are many bulbous, corm-making, rhizomatous or in other ways fleshy-rooted plants that do not really respond happily to being dried off for marketing. They tend to shrivel and lose vitality. Even if you wait till they are more or less dormant before moving them in your own garden and then do it quickly, you'll still, in some cases, not be giving them the optimum treatment, which is to move them green. Move them while their roots are still active so that they can re-establish before becoming dormant.

Lilies are a case in point: the sooner you move them after flowering, the better. If they must be dried off for marketing purposes the period should be kept as short as possible, so that their vulnerable scaly bulbs and perennial basal roots shrivel as little as can be helped. They don't respond well to being treated like hyacinths or tulips, which make right little, tight little bulbs that are genuinely suited to the drying-off technique.

Most of the bulb firms unfortunately handle their lilies in winter, when they've got the mainstream of hyacinth, daffodil and tulip bulbs out of their way. It is a question of their own convenience. It is worth knowing that, in a small way of business, Highland Liliums at Kiltarlity, by Beauly, Inverness-shire, grow their own bulbs and will send early orders out early. Where some of the small bulbs, notably snowdrops and winter aconites, are in question, many suppliers are sending them out in the green; there may be an extra charge, since the cost of packing and postage is greater than sending them dry.

There's a tremendous lot we can do in our own gardens with bulbs that have multiplied into clumps. Which among us can say that we already have as many snowdrops as we want? There's always room for more because they will grow at the foot of shrubs and trees which are heavy with greenery in summer and where we should not normally plant anything at all. It's the same in our borders. I have snowdrops growing between and even out of clumps of border phloxes, and why not? Would we be content to leave the scene to phloxes or the majority of other herbaceous perennials in winter when they've nothing to contribute and the ground is bare? True, when the phloxes need splitting and replanting the bulbs get disturbed, but they can stand that.

So, the minute your snowdrops have faded, probably in early March, dig them up before you forget and do the deed. It's the same with winter aconites, *Eranthis hyemalis*, which many of us find difficult to establish. This is partly a question of finding the right place for them in thin, shady turf, but failures are often on account of their tubers shrivelling when dried off for autumn sales. Erythroniums, the dog's-tooth violets, are notorious in this respect, and you can expect to lose up to 50 per cent of bulbs purchased in the autumn as a matter of course. I never succeeded with the common European *Erythronium dens-canis* until I was given a flowering clump in March 1967. I pulled it into separate units and planted it in various damp, shady places, and it never looked back. Since then I have repeated the exercise.

Most crocus corms do dry off successfully for sales purposes, but if you're increasing your own, spring's the time, when you can see where they are. That applies as well to the autumn as to the spring flowerers; the former make their 'grass' in spring. I had a wonderful time one February, recently, distributing a dense colony of *Crocus speciosus* into rough grass near our horse pond. This is by far the most satisfactory of the true autumn crocuses, with heavily veined, almost blue flowers, bright orange stigmas and a delicious scent. It copes with rough grass magnificently. My quarry was a patch among cyclamen under our bay tree. Here, in weed-free soil (hands-and-knees weeding several times a year) they increase so fast and become so dense as to do themselves no good, after a time. With my bulb planter I took out 400 plugs in four large informal patches and planted from one to half a dozen corms (according to size) in each position. All these came from an area about 18 in. square.

So often this is the best way to set about establishing big colonies. Start off in a small, cheap way with a few bulbs or plants of whatever it is. Grow them in good soil where they can multiply at the rate of knots and then spread them around. The expense is minimal. All you need is a planning temperament which is prepared to wait in some departments while forging ahead in others.

As the details of a job are sometimes helpful, I will mention that these crocuses were planted with the addition of two barrow loads of old potting soil (any good topsoil would have served). The ground into which they were being introduced is horribly clayey sub-soil. One wants to give them a good start. The bulb planter takes out a deep plug. I should specify that this sort of bulb planter is a tool 3 ft long (sometimes they're a little shorter) with a T-handle and a plug-extracting base 2½ in. across. In soft ground you can take out plugs rhythmically at the rate of one every two seconds. Some bulb planters offered for sale are hopelessly unhandy implements not much longer than a trowel and of much too wide a calibre. The sort you want is still offered by Joseph Bentley of Barrow-on-Humber, S. Humberside. They have a huge mail order business and a splendid catalogue of horticultural tools and equipment.

Anyway, I extracted 100 plugs at a time, then followed up with crocuses in one trug, soil in another and a kneeling mat (not to mention my dogs' bed). I break the top 3 in. or so off a plug (this including the best soil and turf fibre), turn it upside down and push it into the bottom of the hole. Then I drop in a large handful of potting soil, press my crocuses into this and firm them with two fingers, adding some more potting soil round their necks.

In the same area I plant more bulbs of *Narcissus asturiensis* (once, more descriptively known as *N. minimus*) each spring after flowering. Again I had my original clump from the kind Northern Irish donor who gave me the dog's-tooth violets in 1967. I split it forthwith and planted it among hyacinths and peonies in a border. Each bulb has made a 6-in. wide clump and I lift a couple of these each year, which gives me between 50 and 100 bulbs. Tiny though its flowers and stature are, like a miniature Lent lily with yellow trumpets, this species is as tough as they come. Most daffodils, with their long stringy foliage, are too impracticable to move in spring, especially in turf, where the grass will already be lengthening by the time the majority of daffodils have faded. But it's a proposition with hoop petticoats – *N. bulbocodium* – which get overcrowded rather easily in borders. So do pheasant's eyes (*N. poeticus* var. *recurvus*) anywhere, becoming barren in more years than not. They scarcely have a resting period, so I should suit your own convenience and lift them in early March while they're still short and manageable. You can move bulbs at any time if you're quick about it. They will suffer only a temporary setback.

The flower of the west wind, *Zephyranthes candida*, has no natural resting period but carries its green rush leaves right through the winter, albeit somewhat frost-bitten. Dried bulb sales therefore obviously don't suit its style, but some general nurseries sell plants green, with their roots attached. In this condition they'll move any time. They carry their white, crocus-like flowers in autumn, so you want to have them well established before then.

So too with the autumn-flowering sternbergias, whose flowers resemble crocuses but are chrome yellow. These do have a complete resting season from May till August, but they don't mind being moved green at any time between autumn and spring, either. You can suit yourself.

The wand flower, *Dierama pulcherrimum*, is evergreen. The roots of young plants are as transparent and fragile as icicles. One raises them from seed, shifts them singly into pots as young as possible and then transplants them, without root disturbance, to their permanent sites where they should start flowering in their third year. Can they really never be disturbed again thereafter, as experts like me are always preaching? Of course they can. They build up chains of persistent old corms over the years (one for each year) in the style of crocosmias (montbretias) and although the young roots at the base of this apparatus may be destroyed on moving them they can regenerate from the old chaps. The plant will sulk for a year but nobody I ever heard tell of died from sulking.

NAKED AND UNASHAMED

'Flowers arising from the bare ground always look a bit odd.' Thus a horticultural journalist who was expressing a widely felt sentiment. Many people to whom I have spoken on the subject have expressed a dislike of colchicums, in particular, because their flowers have the indecency to spring naked from the earth. It is difficult to get at the root of an objection you do not share, but flowers such as these evidently promote a feeling of discomfort and unsuitability in their beholders. *Colchicum autumnale* is not merely known in our language as meadow saffron but also as naked boys and naked ladies. There is a flavour of disapprobation in the adjective.

But it is the very way these elegant flowers spring unattended and unheralded from the (frequently) parched earth that I find so gladdening. W. H. Hudson evidently had similar feelings for these kinds of flower. In *A Crystal Age*, a novel in which he imagines a strange, flawed Utopia, he describes in detail the flowering, in autumn, of the rainbow lilies, at which time the whole community stopped their normal activities simply to enjoy this revitalizing experience. At first the lily appears as 'a small, slender bud, on a round, smooth stem, springing without leaves from the soil'. Then the narrator, who is a stranger in the land, discovers its first, full-blown flower, in shape resembling a tulip but more open, and the colour a most vivid orange-yellow. He is rather disappointed since the name of 'rainbow lily' had led him to expect a many-coloured flower of surpassing beauty. But as more of the lilies gradually open he realizes the reason and why they gave a beauty to the earth which could not be described or imagined. Hudson does describe and imagine it nevertheless.

The flowers were all of one species, 'but in different situations they varied in colour, one colour blending with, or passing by degrees into another, wherever the soil altered its character. Along the valleys, where they first began to bloom, and in all moist situations, the hue was yellow, varying, according to the amount of moisture in different places, from pale primrose to deep orange, this passing again into vivid scarlet and reds of many shades. On the plains the reds prevailed, changing into various purples on hills and mountain slopes; but high on the mountains the colour was blue; and this also had many gradations, from the lower deep cornflower blue to a delicate azure on the summits, resembling that of the forget-me-not and hairbell ... Calm, bright days without a cloud succeeded each other, as if the very elements held the lilies sacred and ventured not to cast any shadow over their mystic splendour.'

Hudson was brought up in the mid-nineteenth-century Argentine and doubtless saw prototypes for his lilies there, perhaps a hippeastrum or zephyranthes. *Zephyranthes grandiflora* is naturalized around hill stations in India and is there known as the thunder lily, because its pink trumpets suddenly rise from the parched ground after a pre-monsoon thunderstorm in early summer. *Z. candida* is another happy settler there which originates from Uruguay.

The nakedness of colchicums evidently gives no cause for concern to numbers of the Dutch, as there is a large-flowered hybrid called 'The Giant' marketed in Holland, to town dwellers in particular. The dry corms are set in dry sand in a bowl and each produces a succession of a dozen or more blooms, these lasting, individually, about three days. No watering or other cultural requirement is necessary. When the display is at an end the material is discarded.

The chief reason for orders of colchicums and autumn-flowering crocuses having to be placed before the end of August is that the corms will otherwise flower prematurely while still in storage or in their packets. Even the most impatient gardener is unlikely to be pleased if he opens a packet of, say, *Crocus speciosus*, to find their blooms, twisted and compressed by their cribbed confinement, peering at him before he has had the chance to plant them.

Like the colchicums, most autumn crocuses flower months ahead of the appearance of their leaves, and this is a particularly happy circumstance if you want to grow them or the meadow saffrons in rough grass. Having enjoyed their flowering, the grass can be given a final cut without risk of damaging crocus or colchicum foliage and before spring-flowering bulbs have pushed through.

One of the most beautiful of these autumn crocuses proclaims its nakedness in its name, *Crocus nudiflorus*. Its flowers are a particularly rich purple, against which its yellow stigmas show up when the blooms open in the sun. I don't think bare earth does make the most flattering background for this flower colour, but it is an excellent colonizer in rough grass, and purple against green is ideal.

Many amaryllids either flower without their leaves or else look as though they wish they could. *Nerine bowdenii* belongs to the latter category. Its flower buds develop in September but are unfortunately surrounded by bundles of lank strap leaves that have been growing all summer. I find it preferable, wherever the leaves show, to make the bulbs flower naked by cutting away the foliage at the turn of August and September, just before the flower buds appear. I have done this for many years on a particular group and it has in no way weakened the bulbs or reduced their freedom of flowering.

Cyrtanthus elatus (syn. *Vallota speciosa*), the Scarborough lily, is an amaryllid that flowers in September. It seems to thrive on neglect, which is a recipe that I find hard to apply. I have seen a 6-in. pot crammed with leafless bulbs bearing thirteen stems, all in bloom. Each stem carried four or five blooms – about three of them out at a time, lasting a week apiece. They expand into wide-mouthed funnels and are of a peculiarly attractive, soft shade of red.

My bulbs never seem to rest completely but always carry some foliage. They have suffered first from the fly whose maggots attack narcissus bulbs – sold to me in some daffodils I grew in a bowl in my greenhouse; then from virus disease. If I could build up a strong stock again, I should love to try them in the garden. William Robinson, in *The English Flower Garden*, claimed that in a warm place, as at the foot of a south wall, vallotas would often thrive better than in a pot, adding, however, that protection is necessary during hard frosts.

One bulb to a pot is how you usually see the large-flowered cultivars of *Hippeastrum*. Although of tropical American origin, they are often marketed as Royal Dutch Amaryllis. I also have my doubts about the blueness of their blood. They look stark and a bit ridiculous seen as singletons, their leaves having barely started to grow, but the bulbs are so huge that it would be asking rather a lot to suggest a group treatment, though think how splendid they would look that way in a tub, decorating a spacious conservatory. The best group of these 'amaryllis' that I ever saw was in a painting by Sonia York. Not only had she given them foliage to flower with but she had put artistic curves into their rod-straight stems!

Amaryllis belladonna is the only true amaryllis; pretty well hardy but free with its pink funnels only in autumns following hot summers and then only when grown at the foot of a hot wall. I love to see its purple scapes rising unadorned from the baked earth.

I suppose our naked flower haters would be thoroughly disapproving of any parasitic flowers that might come their way, since these never have or need foliage. Parasites are all baddies by definition, unless you happen to be one yourself, in which case you become an enterprising entrepreneur.

One of the hardy flower parasites that I most admire (the majority are not hardy in Britain) is *Lathraea clandestina*, which comes from south France and the Iberian peninsula. It parasitizes the roots of willows and the related poplars and also maples. In spring its closely serried tufts of hooded flowers appear, close to the ground, and often make intense splashes of vivid purple over a wide area: for instance along the Backs at Cambridge. They are most noticeable in March, when other more vigorous vegetation like grass has not yet masked them, but their season extends from February till June. At a distance they can easily be mistaken for drifts of crocuses.

You can plant a parasite. I cannot quote the most scientifically efficient method, but I have succeeded simply by transferring lumps of soil full of lathraea roots from a garden where they were established to mine where they were not. Placed near some willow roots, they latched on. These toothworts, as we call them (there is a pasty-faced native species usually found on hazel roots) also seed freely, so you can try your luck with the simplest of all methods: pushing seeds into soft patches of willow-infested mire.

INDEX

Page numbers in *italics* refer to planting plans.

Abelia chinensis 182
    A. × *grandiflora* 182
      A. × *g.* 'Edward Goucher' 182
    A. schumanii 182
Abies koreana 167
Abutilon megapotamicum 36, 180
    A. × *suntense* 124
Acaena 22, 26
    A. novae-zelandiae 153
Acer, Japanese maple 24
    A. palmatum 24, 83
    A. platanoides 67
    A. saccharinum 184
Achillea filipendulina
    'Coronation Gold' 121
Acis autumnalis 113
Achillea 'Coronation Gold' 59
    A. filipendulina 155
    A. 'Moonshine' 59
    A. ptarmica 59
    A. 'Taygetea' 59
Aconitum 57, 112, 121
Adiantum pedatum 169
Aesculus × *carnea* 'Briotii' 67
    A. hippocastanum 67
    A. indica 67
    A. parviflora 95
Agapanthus campanulatus 89
    hardy a. 142
    A. 'Isis' 142
Ageratum 159
    A. Mosaique Group 159

Agrostis canina 150
Ailanthus altissima 40
Ajuga 199
Allangrange, Black Isle, Highland 90
Allium cristophii 121
    A. schoenoprasum var. *sibiricum* 60
alpines 166
Alpine Garden Society show 168
Alstroemeria 157
    A. ligtu hybrids 61
alyssum, rock SEE *Aurinia saxatilis*
Amaryllis SEE ALSO *Hippeastrum*
    A. belladonna 143, 212
Ampelopsis brevipedunculata
    'Elegans' 166
Anchusa 155
Androsace 168
    A. lanuginosa 168
    A. studiosorum (syn. *A. primuloides*)
    168
Anemone × *hybrida* (syn. *A. japonica*) 179
    Japanese a. 112, 129, *129*, 132
      white J. a. 136, 185
angelica tree, Japanese
    SEE *Aralia elata*
annuals 14, 155, 157, 159, 175
Anomatheca laxa 115
Anthemis punctata subsp. *cupaniana*
    26, 134
    A. tinctoria 'Wargrave' 132
Antirrhinum 158
    A. majus, hyacinth-flowered 131
      A.m. Sprite Series 139
      A.m., yellow 155
Apiaceae 190

apple 55
Aralia elata 84
   A. sieboldii SEE *Fatsia japonica*
Araliaceae 94
Arbrex bituminous paint 31
Arbutus unedo 42
Argyranthemum foeniculaceum 25
   A. frutescens 25
Armillaria mellea 55
Artemisia arborescens 180, 185, 190
Arum italicum 'Marmoratum'
     (syn. *A.i.* 'Pictum') 113
   wild arum 58
arum lily, white
   SEE *Zantedeschia aethiopica*
ash SEE *Fraxinus*
   common a. SEE *F. excelsior*
   manna a. SEE *F. ornus*
Asplenium trichomanes 116
Aster 157
   A. amellus 59
   A. × *frikartii* 59, 179
Asteraceae 44
aster, China SEE *Callistephus chinensis*
Astilbe 'Fanal' 121
Astrantia major
   'Sunningdale Variegated' 204
Athyrium 61
Aubrieta 26, 117
Aucuba japonica 'Crotonifolia' 109
   A.j. 'Picturata' 109
Aurinia saxatilis 'Citrina' 117
azalea SEE *Rhododendron*
   Kurume a. 103
Azara microphylla 44

Ballota pseudodictamnus 26,
   185
balsam 91
bamboos 146
Baptisia australis 60, 89
Barbarea vulgaris 'Variegata' 133
Barilli & Biagi, Bologna, Italy
   (seedsmen) 65
bay laurel SEE *Laurus nobilis*
Bean, William Jackson, *Trees and*
   Shrubs Hardy in the British Isles
   37, 81, 82, 181
Bedgebury, National Pinetum, Kent 165
bedding plants, tender 175
beech 32
   purple b. 65
   weeping b. 66
Bellis perennis Monstrosum Group 127
   B.p. Pomponette Series 127
Ben Lawers, Perthshire 201
benomyl fungicide 104, 139
Bentley, Joseph, Barrow-on-Humber,
   Lincs. (garden tools) 207
Berberis 35
   B. temolaica 40, 90
Bergenia (syn. *Megasea*) 61, 107, 110,
   154, 157
   B. 'Ballawley' 121
   B. ciliata 110
   B. cordifolia 121
   B. 'Morgenrot' 61
   B. purpurascens 110
biennials 14
bindweed 58
birch, weeping 66

bitter cress, hairy
    SEE *Cardamine hirsuta*
blood, fish and bone fertilizer 49
Bloom, Alan 153
blue-eyed grass
    SEE *Sisyrinchium angustifolium*
bluebell SEE *Hyacinthoides*
    English b. SEE *H. non-scripta*
bonsai 51
border plans 145–9
Botrytis cinerea 22
Bougainvillea 120
box SEE *Buxus*
Brachyglottis rotundifolia 44
    B. laxifolia 189
    B. greyi 189
    B. 'Sunshine' 44, 189
bramble 153
broom SEE *Cytisus, Genista*
    Mount Etna b. SEE *Genista aetnensis*
bryony, black SEE *Tamus communis*
buckeye, dwarf SEE *Aesculus parviflora*
Buddleja 24, 57, 84, 91
    B. alternifolia 'Argentea' 90
    B. davidii 40, 48
        B.d. 'Harlequin' 103
        B.d. 'Royal Red' *101*, 103
    B. fallowiana 'Alba' 84, 171
    B. 'Glasnevin Hybrid' 84
    B. 'Lochinch' 84
bugloss, annual
    SEE *Echium plantagineum*
bulb planter 206
bulbs to plant green 205–9
bullfinches 45, 143
Bupleurum fruticosum 118
Burford House, Tenbury Wells, Worcs.
    88–90, 95, 194
Busch, Wilhelm 200
Buxus 20

Calceolaria mexicana 201
Callistemon 34
    C. subulatus 34
Callistephus chinensis 159
    C.c. Bouquet Powderpuffs Group
        130
    C.c. Princess Series 131
Camellia 21, 24, 38
    C. japonica 109
        C.j. 'Adolphe Audusson' 109
        C.j. 'Alba Simplex' 109
    C. saluenensis 109
    C. × *williamsii* 109
        C. × *w.* 'Donation' 38
        C. × *w.* 'J.C. Williams' 109
Cameron, Major & Mrs Allan John 90
Campanula 26
    C. cochlearifolia (syn. *C. pusilla*) 116
    C. lactiflora 90, 121, 143, 155
    C. portenschlagiana
        (syn. *C. muralis*) 26, 115
    C. poscharskyana 26, 61, 115
        C.p. 'Stella' 61, 116
    C. pyramidalis 155
Campsis grandiflora (syn. *C. chinensis*)
    142
candytuft SEE *Iberis*
Canna 155, 156, 157
Cape daisy SEE *Felicia amelloides*
capsids 157
captan fungicide 22, 29
Cardamine hirsuta 58
cardoon 173
carnation rust disease 131
carrot 190
Caryopteris 21, 86
    C. × *clandonensis* 97, 99, *100*
        C. × *c.* 'Arthur Simmonds' 97
        C. × *c.* 'Ferndown' 86, 97
        C. × *c.* 'Heavenly Blue' 86, 97
        C. × *c.* 'Kew Blue' 86, 97, 99
Cassinia leptophylla subsp. *fulvida* 96

Castanea sativa 68, 107
Catalpa, golden 40
Ceanothus 23, 25, 36, 86
    C. *dentatus* var. *floribundus* 36
    C. × *delileanus* 'Gloire de Versailles'
        86, *100*, 101, 102
    C. *divergens* 202
    C. *impressus* 36
    C. *thyrsiflorus* var. *repens* 118
celandine 58
celery 190
Celosia 156, 158
Centaurea gymnocarpa 25, 185, 189
Centranthus ruber 119
Ceratostigma plumbaginoides 116
    C. *willmottianum* 86
Cestrum parqui 181
Chaenomeles 46
    C. *speciosa* 36, 45
Chamaecyparis lawsoniana 33
    C. *nootkatensis* SEE *Xanthocyparis n.*
Chatto, Beth 114, 204
Cheiranthus SEE *Erysimum*
Chelsea Physic Garden, London 194
cherry 45
    'Kanzan' c. SEE *Prunus* 'K.'
    morello c. 36
    winter c. SEE *Prunus*
    × *subhirtella* 'Autumnalis' cherry
        laurel 45
chestnut, horse
    SEE *Aesculus hippocastanum*
    Indian h.c. SEE *Aesculus indica*
chestnut, sweet or Spanish
    SEE *Castanea sativa*
Chimonanthus praecox 39, 172, 184
Chinese pagoda tree
    SEE *Styphnolobium japonicum*
chives, giant SEE *Allium schoenoprasum*
    var. sibiricum
Choisya ternata 21, 22, 43, *101*, 102, 110
Chondrostereum purpureum 41, 45

Christmas rose SEE *Helleborus niger*
Chrysanthemum 120
    C. *ptarmiciflorum* SEE *Tanacetum p.*
Cineraria maritima
    SEE *Senecio cineraria*
Cistus 25, 88, 186
    C. *albidus* 186
    C. × *cyprius* 23, 36
    C. × *hybridus*
        (syn. C. × *corbariensis*) 36, 90
    C. × *purpureus* 89
    C. × *skanbergii* 118, 186
Clematis 24, 26, 89, 158
    C. *alpina* 'Pamela Jackman' *100*
    C. 'Bill MacKenzie' 181
    C. 'Columbine' 42
    C. 'Comtesse de Bouchaud' 43
    C. 'Durandii' 119
    C. *flammula* 119
    C. 'Gravetye Beauty' *101*, 104
    C. 'Huldine' 42
    C. 'Jackmanii Superba' 43
    C. × *jouiniana* 181
    C. 'Lady Betty Balfour' 142
    C. *montana* 42
    C. 'Nelly Moser' 43
    C. 'Perle d'Azur' 42
    C. 'Praecox' (syn. C. × *jouiniana*
        'P'.) 119, 181
    C. *tangutica* 181
    C. *tibetana* subsp. *vernayi* 181
    C. *tubulosa* (syn. C. *davidiana*) 155
    C. *viticella* subsp. *campaniflora* 181
    C. 'White Moth' 99, *100*
Clerodendrum bungei
    (syn. C. *foetidum*) 93
    C. *trichotomum* 95
    C.t. var. *fargesii* 95
cockscomb SEE *Celosia*
Colchicum 115, 177, 210
    C. *autumnale* 209
    C. *speciosum* 'Album' 113

C. 'The Giant' 210
compost, cutting, John Innes 18
    John Innes No. 1 c. 135, 138
    J.I. No. 3 c. 52, 133, 187
concrete paving 151
conifer 21, 25
Convolvulus althaeoides
    subsp. *tenuissimus* 186
    C. cneorum 24, 118, 184, 186
    C. sabatius (syn. *C. mauritanicus*)
    25, 178
copper fungicide 131
coral spot fungus
    SEE *Nectria cinnabarina*
Coreopsis 155
    C. 'Badengold' 60
    C. verticillata 60
Cornus kousa var. *chinensis* 53
Coronilla valentina subsp. *glauca* 52
Cortaderia richardii 198
    C. selloana 195–9
    C.s. 'Monstrosa' 198
    C.s. 'Pumila' 198
    C.s. 'Rosea' 197
    C.s. 'Sunningdale Silver' 197
    C.s. 'Violacea' 197
Corydalis lutea 116
    C. ochroleuca 116
Corylopsis 37
Corylus maxima 'Purpurea' 40
Cotinus coggygria 35, 40, 90
Cotoneaster 46
    C. horizontalis 46, 99, *100*, 117
    C. microphyllus 152
cow parsley 58, 164
crab apple SEE *Malus*
    Siberian c. SEE *M. baccata*
cranesbill SEE *Geranium*
    bloody c. SEE *G. sanguineum*
crazy paving 151
Crinum × *powellii* 178

Crocosmia 209
    C. masoniorum 142
Crocus 114, 207, 210
    C. chrysanthus 'Blue Pearl' 122
      C.c. 'Cream Beauty' 122
      C.c. 'Ladykiller' 122
    C. flavus (syn. *C. aureus*) 122
    C. nudiflorus 211
    C. sieberi 'Violet Queen' 122
    C. speciosus 113, 207, 210
    C. tommasinianus 113
      C.t. 'Whitewell Purple' 113
Cuphea cyanea 25, 176
    C. ignea 25, 176
Cupressus macrocarpa 33, 34
× *Cuprocyparis leylandii*
    (syn. × *Cupressocyparis l.*) 33, 34
currant, flowering 30
cuttings 17–30
    c. of evergreens 18–19
    hardwood c. 17–21
    internodal (single-node) c. 27
    root c. 29
cutting back 30, 57
Cyclamen 115, 166, 206
    C. coum 112
    C. hederifolium 112
    C. repandum 112
cypress 21
    Lawson c.
      SEE *Chamaecyparis lawsoniana*
    Leyland c.
      SEE × *Cuprocyparis leylandii*
Cymbalaria muralis 115, 175
Cyrtanthus elatus
    (syn. *Vallota speciosa*) 211
Cystopteris fragilis 116
Cytisus 21, 24, 36
    C. 'Hollandia' 23, *100*
    C. × *kewensis* 36, 119
    C. × *praecox* 'Goldspeer' 201

Dactylorhiza 41, 114
   D. foliosa 114
   D. maculata 114
daffodil SEE *Narcissus*
   hoop-petticoat d. SEE *N. bulbocodium*
Dahlia 155, 156, 157
   D. Coltness Series 136, 140
      D. (C.S.) 'Coltness White' 136
   D. ' Redskin Group 135
daisy SEE *Bellis perennis*
daisy bush SEE *Olearia*
Danaë racemosa 108
Daphne 21, 24
   D. mezereum 110
   D. laureola 110
      D.l. subsp. *philippi* 110
   D. odora 21
   D. pontica 21, 110
   D. tangutica 21
      D.t. Retusa Group 21
Dasineura affinis 203
Davenport-Jones, Hilda 124
Davidia involucrata 53
daylily SEE *Hemerocallis*
Delphinium 60, 90, 155, 157
Dendromecon 25
Deutzia 18, 40
   D. × *rosea* 'Carminea' 99, *100*, 201
Dianthus 22
   D. Allwoodii pinks 178
   D. Ballet Group 135
   D. 'Doris' 178
   D. Highland Group 135
Dierama pulcherrimum 208
Dimorphotheca barberae
   SEE *Osteospermum jucundum*
Dipelta 40
dog's-tooth violet SEE *Erythronium*
dogwood SEE *Cornus*
Doronicum 57
Dorotheanthus bellidiformis 130
Dover, Michael 13

Dracunculus vulgaris
   (dragon arum) 169
drainage 50
Drake, Jack, Inshriach Nursery 177
dropwort, water SEE *Oenanthe crocata*
Drummond Castle, nr Stirling 188
dry walls 115
Dryopteris 61
Dumpton Park, Broadstairs, Kent 67
Dutch elm disease 63

Echium plantagineum
   'Blue Bedder' 132
Edinburgh, Royal Botanic Garden 205
eelworm 53
Elaeagnus angustifolia 184
   E. pungens 'Maculata' 19, 23, 24,
      83, 99, *100*, 170
elder SEE *Sambucus*
elm SEE *Ulmus*
   English e. SEE *U. procera*
   wych e. SEE *U. glabra*
Epilobium glabellum 176
Eranthis hiemalis 113, 205
Erica 35, 91
   E. carnea 'King George' 35
   E. erigena (syn. *E.mediterranea*) 35
Erigeron glaucus 175
   E. g. 'Elsie' 176
   E. g. 'Elstead Pink' 176
   E. g. 'Ernest Ladhams' 176
   E. karvinskianus (syn. *E.
      mucronatus*) 115, 175
Erinus alpinus 116
Eryngium 190
   E. agavifolium (*E. bromeliifolium*
      misapplied) 194
   E. alpinum 192
      E.a. 'Improved' 192
   E. amethystinum 192
   E. bourgatii 192
   E. campestre 183

E. eburneum (syn. *E. paniculatum*) 194

E. giganteum 193

E. maritimum 191, 193

E. × *oliverianum* 155, 191, 192

E. pandanifolium 121, 193, 194, 195

E. planum 192

   E.p. 'Blauer Zwerg' 192

E. proteiflorum (*E.* 'Delaroux') 54, 195

E. serra 194

E. × *tripartitum* 192

E. variifolium 192

E. 'Violetta

New World e. 193

Old World e. 191, 195

Erysimum 22, 139

   E. 'Bowles's Mauve' 135

   E. cheiri 'Carmine King' 129

      E.c. 'Cloth of Gold' 130

      E.c., double-flowered 132

      E.c. 'Fire King' 130

      E.c. 'Harpur Crewe' 125

      E.c. 'Ivory White' (syn. *E.c.* 'White Dame') 130

      E.c. Persian Carpet Group 129

      E.c. 'Primrose Monarch' 129, 130

      E.c. 'Purple Queen' (syn. *E.c.* 'Ruby Gem') 130

   E. linifolium 131

Erythrina crista-galli 98

   E.c.-g. 'Compacta' 98

Erythronium 114, 205, 208

   E. dens-canis 205

   E. revolutum 'Album' 114

Escallonia 20, 42

   E. 'Apple Blossom' 42

   E. bifida (syn. *E. montevidensis*) 92

   E. 'Iveyi' 23, 42, 92

   E. × *langleyensis* 42

Eucalyptus 34

   E. gunnii 34, 90

E. perriniana 35

Eucryphia 25, 91

   E. glutinosa 92

   E. × *intermedia* 92

      E. × *i.* 'Rostrevor' 92

   E. × *nymansensis* 53

      E. × *n.* 'Nymansay' 25, 91, 92

Euphorbia 124

   E. amygdaloides var. *robbiae* 108, 116, 119

   E. characias 19, 25

      E.c. subsp. *wulfenii* 25, 108, 119, 121

   E. griffithii 124

      E. g. 'Fireglow' 124

      E. g. 'Dixter' 124

   E. myrsinites 201

   E. palustris 124

   E. polychroma 119, 124

   E. seguieriana subsp. *niciciana* 142, 179

evening primrose SEE *Oenothera*

Exbury Gardens, Beaulieu, Hants. 122

Fallopia baldschuanica 94

Farrer, Reginald 167–170

   The English Rock Garden 167

Fatsia japonica 41, 109

Felicia amelloides 155

   F. pappei 25

fennel 190

   purple-leaved f. 127

ferns 41, 61, 111, 116

   beech f. SEE *Phegopteris connectilis*

   bladder f. SEE *Cystopteris fragilis*

   maidenhair f. SEE *Adiantum*

   oak f. SEE *Gymnocarpium dryopteris*

fig 53

   'Black Ischia' f. 53

   'Brown Turkey' f. 53

   'Brunswick' f. 53

filbert SEE *Corylus maxima*

finger blight 166
Fish, Margery 126, 152
flower of the west wind
    SEE *Zephyranthes candida*
forget-me-not SEE *Myosotis*
    water f.-m.-n. SEE *M. scorpioides*
Forsythia 30, 120
    F. × *intermedia* 'Lynwood' 99, *100*
        F.×*i.* 'Spectabilis' 98, *100*
Fothergilla monticola 37
foxglove 114
Fraxinus 64, 65, 153
    F. excelsior 65, 67
        F.e. 'Pendula' 66
    F. mariesii 67
    F. ornus 66
Fuchsia 25, 57, 85, 91, 157, 182
    F. 'Corallina' 117
    F. 'Dollar Princess' 166
    F. 'Eva Boerg' 182
    F. 'Genii' 128
    F. 'Granny's Weeper' 117
    F. 'Lena' 117, 182
    F. magellanica 'Versicolor' (syn.
        F. gracilis 'Tricolor') *100*, 102
    F. 'Marinka' 117
    F. 'Mrs Popple' 128, 182
fumitory, yellow SEE *Corydalis lutea*
fungicides 22, 190

Gaillardia 178
Galanthus platyphyllus 114, 204
gall midge, violet 203
Galtonia candicans 85
Garrya 25
    G. elliptica 14, 96
Gazania 22, 25, 186
Genista aetnensis 141
    G. canariensis (florists' g.) 52
    G. hispanica 36
    G. lydia *101*, 105, 119
Gentiana acaulis 127

Geranium (Cinereum Group)
    'Ballerina' 177
    G. endressii 177
    G. macrorrhizum 59
    G. × *oxonianum* 'A.T. Johnson' 177
        G. × *o.* 'Wargrave' 177
    G. psilostemon 121
    G. 'Russell Prichard' 177
    G. sanguineum 177
        G.s. var. *lancastriense* 177
    G. wallichianum 'Buxton's Variety'
        (syn. *G.w.* 'Buxton's Blue') 177
geranium SEE ALSO *Pelargonium*
Geum borisii 140
    G. × *heldreichii* 'Superbum' 201
    G. rivale 126
Gladiolus 156, 158
    Butterfly g. 158
    G. communis subsp. *byzantinus* 158
    Grandiflorus g. 158
Glasnevin Botanic garden, Dublin 66
goose grass 58
gorse SEE *Ulex europaeus*
Gould, Arthur 33
grape hyacinth SEE *Muscari*
grasses 61
    Bowles's golden grass
    114, 204
gravel 151
Great Dixter, Northiam, East Sussex,
    Long Border 33, 40, 42, 55, 120,
        157, 158, 188
    rose garden 91
grey-leaved plants 184–90
    propagation 189–90
grey mould SEE *Botrytis cinerea*
Griselinia littoralis 21
ground cover 59
ground elder 58
Gymnocarpium dryopteris 116
Gypsophila paniculata 156

Halliwell, Brian 127
Hamamelidaceae 37
Hamamelis 45
   H. mollis 37
handkerchief tree
   SEE *Davidia involuvcrata*
Hay, Roy 159
heather SEE *Erica*
Hebe 19, 21, 38, 91
   H. 'Autumn Glory' *101*, 104
   H. 'Gauntlettii' 38
   H. 'Glaucophylla Variegata' 128
   H. 'Great Orme' 38
   H. 'Midsummer Beauty' 181
   H. 'Mrs Winder' 738, *101*, 104
   H. speciosa 181
hedges 31, 32, 49–51
Hedychium 54
Helenium 155
Helianthemum 26, 117
Helianthus 155
Helichrysum microphyllum
   SEE *Plecostachys serpyllifolia*
   H. petiolare 25, 176, 187
   H. splendidum 23, 44, 85, 188
Heliopsis 192
heliotrope 'Marine' 133
   h. 'Regale' 139
Helleborus (hellebore)
   H. argutifolius (syn. *H. lividus*
     subsp. *corsicus*) 109
   H. foetidus (stinking h.) 108, 169
   H. × *hybridus* (syn. *H. orientalis*
     hybrids) 113
   H. lividus 109
   H. niger 113
     H.n. var. *altifolius* 113
   H. orientalis subsp. *abchasicus*
     Early Purple Group 11, 113
Hemerocallis 60
   H. fulva, double 156
   H. 'Helios' 86

   H. lilioasphodelus (syn. *H. flava*) 60
hemlock 190
herbaceous perennials, replanting 59
Heuchera 61
× *Heucherella* 61
Hibiscus sinosyriacus 35
   H. syriacus 35
     H.s. 'Coelestis' 97
     H.s. 'Oiseau Bleu' 97
Highland Liliums, Kiltarlity, by Beauly,
   Inverness-shire 205
Hillier, Sir Harold George Knight 194
Hillier Arboretum (Sir Harold Hillier
   Gardens), Romsey, Hants. 122
Hilliers of Winchester, Hants.
   (nursery) 65, 182
Hippeastrum 52, 210, 211
hogweed 190
Hoheria glabrata 95
   H. lyallii 95
   H. populnea 95
     H.p. 'Alba Variegata' 95
     H.p. 'Foliis Purpureis' 95
   H. sexstylosa 95
holly SEE *Ilex*
hollyhock 156
honesty SEE *Lunaria annua*
honey fungus SEE *Armillaria mellea*
honeysuckle SEE *Lonicera*
   Dutch h. SEE *L. periclymenum*
     'Serotina'
hormone rooting powder/dip 22
hornbeam 32
hortensia SEE *Hydrangea macrophylla*
Hosta 41, 54, 60
   H. crispula 60
   H. 'Honeybells' 60, 171, 172
   H. plantaginea 171
   H. rectifolia 85
   H. 'Royal Standard' 172
   H. ventricosa 85
   H. undulata 119

Hudson, William Henry 195
   A Crystal Age 209
   The Naturalist in La Plata 196
hyacinth 91, 205
Hyacinthoides hispanica
   (syn. *Scilla campanulata*) 114
   H. non-scripta 114
Hydrangea 19, 24, 57, 91, 141, 155
   H. anomala subsp. *petiolaris* 90, 108
   H. arborescens 48, 49
   H. macrophylla 47, 85, 87
     H.m. 'Mme Emile Mouillère' 183
   H. paniculata 24, 48, 85
     H.p. 'Floribunda' 48
     H.p. 'Grandiflora' 48, *101*, 102
     H.p. 'Praecox' 49
     H.p. 'Tardiva' 48
   H. 'Preziosa' *101*, 102, 183
   h. pruning 47
Hypericum 91
   H. 'Hidcote' 85, 86
   H. × *inodorum* 'Elstead' 8
   H. 'Rowallane' 86

Iberis sempervirens 117
Ilex 19, 20, 24
   I. × *altaclerensis* 'Golden King'
     83, 171
   I. aquifolium 32
Indigofera 57
   I. heterantha (syns. *I. dosua*,
     I. gerardiana) 85
   I. kirilowii 97
Ingram, Collingwood 'Cherry'
   38, 117, 126
International Plant Propagators'
   Society 24, 27
Inverewe Garden, Poolewe, Ross-shire
   122
Iris 61
   Dutch i. 125, 134, 135, 156
   I. foetidissima 110

I. 'Ideal' 134, 135
I. japonica 110
I. pallida 'Argentea Variegata' 89
I. 'Purple Sensation' 136, 137
I. unguicularis (syn. *I. stylosa*)
   112, 143, 184
Itea ilicifolia 96
ivy 49

Jasminum (jasmine)
   J. nudiflorum (winter j.) 36
   summer j. 24
Jekyll, Gertrude 153–60
   Colour in the Flower Garden 153, 155
   Wood and Garden 153, 160
Jerusalem sage SEE *Phlomis fruticosa*
Jiffy No 7 pots 25
Jobling, John 34
jonquil SEE *Narcissus*
Judas tree 29
jumping Jesus SEE *Cardamine hirsuta*
Juniperus (juniper) 21, 88
   J. sabina 'Tamariscifolia' 119
   Pfitzer's j. 89

Kerria japonica 45
Kniphofia 156
   K. 'Modesta' 142
Kolkwitzia amabilis 40, *101*, 104
   K.a. 'Pink Cloud' 104

labels 163
Lamium maculatum 54, 107
Lapeirousia laxa SEE *Anomatheca l.*
Lathraea clandestine 212
Lathyrus latifolius 118
   L. rotundifolius 118
Laurus nobilis 20, 143, 207
Lavandula (lavender) 20, 22, 24, 37,
   127, 167, 190
Lavatera × *clementii* 179
   L. olbia 85

lawns 164
lenacil weedkiller 140
Lent lily SEE *Narcissus pseudonarcissus*
Leucanthemum hosmariense
    SEE *Rhodanthemum h.*
    L. × *superbum* 'Esther Read' 178
Leucojum aestivum 124
    L.a. 'Gravetye Giant' 124
    L. autumnale SEE *Acis autumnalis*
Ligustrum 93, 171
    L. lucidum 93
    L. quihoui 93
lilac SEE *Syringa*
Lilium (lily) 157, 167, 205
    L. auratum 155
    L. lancifolium (syn. *L. tigrinum*) 156
    L. longiflorum 155
lime SEE *Tilia*
    European white l. SEE *T. tomentosa*
    small-leaved l. SEE *T. cordata*
    weeping silver l. SEE *T.* 'Petiolaris'
Lithodora diffusa 'Heavenly Blue'
    (syn. *Lithospermum diffusum*
    'H.B.') 26
    L. oleifolia (syn. *Lithospermum
    oleifolium*) 170
Lithospermum diffusum
    SEE *Lithodora diffusa*
    L.oleifolium SEE *Lithodora oleifolia*
Livingstone daisy
    SEE *Dorotheanthus bellidiformis*
Lloyd, Christopher 9–11
Lloyd, Daisy 9
Lloyd, Nathaniel 9
Lobelia 155
    L. cardinalis 185
    perennial l. 60, 157
Lonicera 20, 43
    L. fragrantissima 43
    L. japonica 'Halliana' 141
    L. nitida 33
        L.n. 'Baggesen's Gold' 33, 57

    L. periclymenum 'Serotina' 141
    L. × *purpusii* 20, 43
Loudon, John Claudius 66
Luma apiculata 95
Lunaria annua 123
lungwort SEE *Pulmonaria*
lupin 60, 139
Lychnis chalcedonica 90, 156
    L. coronaria 186
Lysichiton americanus 123

Magnolia 24
    M. campbelii subsp. *mollicomata* 123
    M. grandiflora 183
        M.g. 'Ferruginea' 183
        M.g. 'Goliath' 183
        M.g. 'Maryland' 183
    M. kobus 53
Mahonia 19, 44
    M. japonica 23, 26, 44, 83, 110, 184
    M. lomariifolia 44, 84, 110
    M. × *media* 'Buckland' 44, 110
        M. × *m.* 'Charity' 43, 110
        M. × *m.* 'Lionel Fortescue' 44, 110
        M. × *m.* 'Winter Sun' 44
    M. × *wagneri* 'Undulata' 44, 83
Malus 79–82
    M. × *atrosanguinea* 82
        M. × *a.* 'Gorgeous' 81
    M. baccata 80
    M. baccata var. *mandshurica* 80
    M. 'Crittenden' 80
    M. 'Dartmouth' 81
    M. floribunda 81
    M. × *gloriosa* 'Echtermeyer' 81
    M. hupehensis 81
    M. 'John Downie' 81
    M. × *moerlandsii* 'Liset' 8
        M. × *m.* 'Profusion' 81
    M. 'Neville Copeman' 80
    M. niedzwetzkyana 80

M. × *robusta* 79
   M.× *r.* 'Aldenhamensis' 80
   M. × *r.* 'Eleyi' 80
   M.× *r.* 'Lemoinei' 80
   M.× *r.* 'Red Sentinel' 80
M. sargentii 82
M. × *schiedeckeri* 'Red Jade' 80
M. 'Simcoe' 81
M. trilobata 82
M. × *zumi* 'Golden Hornet' 81
manure, organic 105
maple SEE *Acer*
   Norway m. SEE *A. platanoides*
   silver m. SEE *A. saccharinum*
marjoram SEE *Origanum vulgare*
marigold, African 155
Marston, Elizabeth 27–9
Mason, Maurice 29
meadow gardening 164
meadow saffron
   SEE *Colchicum autumnale*
meadowsweet, double 155
Melianthus major 40
Men of the Trees, The 66
Mertensia virginica 111
Mesembryanthemum criniflorum
   SEE *Dorotheanthus bellidiformis*
Mexican orange SEE *Choisya ternata*
Michaelmas daisy 59, 185
Mimulus glutinosus 25, 176
   M. puniceus 176
Miscanthus 146
Miss Willmott's ghost
   SEE *Eryngium giganteum*
mist unit 22
Mitchell, Alan Fyson 66, 69
mixed borders 82–7
Monarda 157, 158
   M. 'Beauty of Cobham' 133
monkshood SEE *Aconitum*
montbretia 54
   SEE ALSO *Crocosmia*

Mörike, Eduard Friedrich, *Heimweh*
   200
mulberry 57
Muscari armeniacum 'Blue Spike' 127
Myosotis 199
   M. alpestris 201
   M. palustris 200
   M. scorpioides (syn. *M. palustris*) 200
   M.s. 'Mermaid' 200
Myrtus (myrtle) 25, 34
   M. communis (common m.) 21, 96
   M. luma SEE *Luma apiculata*

naked boys SEE *Colchicum autumnale*
naked ladies, SEE *Colchicum autumnale*
Nandina domestica 97
Narcissus 114, 150, 205, 208
   N. asturiensis (syn. *N. minimus*)
     114, 208
   N. bulbocodium 114, 208
     N.b. var. *citrinus* 114
   N. cyclamineus 112
   N. poeticus var. *recurvus*
     (pheasant's eye n.) 126, 208
   N. pseudonarcissus 114
   N. 'Tittle-tattle' 125
narcissus fly 211
nasturtium 156
   n. Gleam Series 138
   n. (G.S.) Golden Gleam' 138
   n. Whirlybird Series 138
   n. (W.S.) 'Whirlybird Gold' 138
National Gardens Scheme 161
National Trust 166
Nectria cinnabarina 39, 95
Nerine bowdenii 185, 211
Nothofagus 78
   N. dombeyi 78
Nottingham University School of
   Agriculture 27

oak SEE *Quercus*
    holm o. SEE *Q. ilex*
    pin o. SEE *Q. palustris*
    scarlet o. SEE *Q. coccinea*
    Turkey o. SEE *Q. cerris*
Oakover Nurseries, Ashford, Kent 64
Oenanthe crocata 123
Oenothera 176
    O. macrocarpa (syn. *O. missouriensis*)
    176
Olearia 21, 96
    O. avicenniifolia 96
    O. × *haastii* 96
    O. macrodonta 44, 146
    O. × *scilloniensis* 201
    O. solandri 96
oleaster SEE *Elaeagnus angustifolia*
Omphalodes cappadocica 111, 125
    O. luciliae 170
Onopordum acanthium 174
Opuntia 174
orchid, hardy SEE *Orchis, Dactylorhiza*
    wild o. 164
Orchis 114
    O. maculata SEE *Dactylorhiza m.*
Origanum vulgare 'Aureum' 127
Ornithogalum nutans 114
Osmanthus delavayi 21, 23, 83
Osteospermum 25
    O. jucundum 176
    O. 'White Pim' 176
Oteley Garden, Ellesmere, Shropshire
    151
Oxalis corniculata var. *atropurpurea*
    91
Ozothamnus 45
    O. ledifolius 125
    O. rosmarinifolius 45

Paeonia 60
    P. lactiflora hybrids 128
pampas grass SEE *Cortaderia selloana*

pansy SEE *Viola*
parasitic plants 212
Parrotia 37
parsley 190
    p. 'Bravour' 135
parsnip 190
Passiflora caerulea
    (common passion flower) 180
paths 149–52
Paulownia tomentosa 40
pea, everlasting SEE *Lathyrus*
peach 36, 45
pear SEE *Pyrus*
Pelargonium 25, 121
    P., Ivy-leaved 121
Penstemon 25, 57, 156, 179
    P. 'Andenken an Friedrich Hahn'
        (syn. *P.* 'Garnet') 179
    P. 'Drinkstone Red' 134, 179
    P. 'Evelyn' 179
    P. 'Sour Grapes' 179
peony SEE *Paeonia*
perennials, short-lived 155
    tender p. 155
periwinkle SEE *Vinca*
Pernet-Ducher, Joseph 88
Persicaria 178
    P. affinis 'Darjeeling Red 171
        P.a. 'Donald Lowndes' 171, 178
    P. amplexicaulis 'Atrosanguinea' 178
    P. milletii 178
Petunia 'Blue Dandy' 134
Pieris 'Forest Flame' *100*
Phegopteris connectilis 116
Philadelphus 18, 30, 40
    P. 'Belle Etoile' *101*, 101
    P. coronarius 'Aureus' 84
    P. 'Virginal' 100, *101*, 101
Phlomis 186
    P. chrysophylla 43
    P. drummondii Cecily Group 137
    P. fruticosa 23, 43, 85, 188

Phlox 24, 57, 112, 141, 143, 156, 158, 163, 205
    P. 'Mia Ruys' 177
photographers 164
Phygelius aequalis 180
    P. capensis 179
Pileostegia viburnoides 94
Piptanthus nepalensis
    (syn. *P. laburnifolius*) 123
Pittosporum tenuifolium 21, 42
plane 107
planting 15
Plecostachys serpyllifolia 25
plum 45
Plumbago auriculata (syn. *P. capensis*) 155
Polygonum affine
    SEE *Persicaria affinis*
    P. baldschuanicum
    SEE *Fallopia baldschuanica*
Polypodium (polypody) 111
    P. interjectum 'Cornubiense' 111
    P. vulgare (common p.) 111, 117
Polystichum 24, 61
pomegranate SEE *Punica granatum*
poppy 146, 167
    opium p. 167
    Oriental p. 120, 156
Portugal laurel 45
Potentilla, shrubby 46, 85, 91, 97
    P. fruticosa 'Elizabeth'
        (syn. *P.f.* 'Arbuscula') 99, *100*
    P. fruticosa 'Jackman's Variety' 182
    P.f. 'Katharine Dykes' 182
    P.f. 'Red Ace' 183
    P.f. 'Sunset' 183
    P.f. 'Tangerine' 183
prickly pear 174
primrose 113
Primula denticulata 123
privet SEE *Ligustrum*

pruning 15, 30–49
Prunus 31, 45
    P. glandulosa 'Alba Plena' 126
    P. × *subhirtella* 'Autumnalis' 45
Puddle, Charles 92
Pulmonaria longifolia 111
Punica granatum 35
Pyracantha 46
Pyrus 36
    P. amygdaliformis 64

Quercus 32, 64
    Q. cerris 65
    Q. coccinea 65
        Q.c. 'Splendens' 65
    Q. ilex 32
    Q. palustris 65
    Q. rubra (syn. *Q. borealis*) 65
        Q.r. 'Aurea' 65
Quihou, Antoine 93
quince, Japanese
    SEE *Chaenomeles speciosa*

Ranunculus flammula 200
    R. gramineus 125, 136
red cedar, Western SEE *Thuja plicata*
red hot poker SEE *Kniphofia*
replanting herbaceous perennials 59
Rhodanthemum hosmariense 176
Rhodiola rosea 201
Rhododendron 24, 37, 45, 53, 64, 83, 109, 120, 121
    R. arborescens 83
    R. arboreum 109
    R. barbatum 37
    R. bureavii 109
    R. campanulatum 109
    R. cinnabarinum 110
        R.c. subsp. *xanthocodon*
            Concatenans Group 109
    R. fortunei subsp. *discolor* 37
    R. 'Hinomayo' *101*, 103

R. mallotum 109
R. ponticum 55
R. 'Tessa' 122
R. thomsonii 37
R. viscosum 83
R. yakushimanum 109
Rhus cotinus SEE *Cotinus coggygria*
   R. glabra 'Laciniata' 40
Ribes 40
Robinson, William,
   The English Flower Garden 168, 211
root pruning 51
rooting aids 22
rooting hormones 22
Rosa (rose) 19, 20, 24, 26–30, 45,
   46, 57, 87–91, 143, 146, 156, 157
   Alba r. 90
   R. × alba (white rose of York) 90
   R. 'Albertine' 172
   R. 'Allotria' 27, 28
   R. 'Belle de Crécy' 46
   R. 'Blanche Double de Coubert' 46
   R. 'Céleste' 90
   Centifolia r. 89
   R. 'Chinatown' 46
   Damask r. 89
   R. 'De Meaux' 89
   R. 'De Resht' 89
   R. 'Duc de Guiche' 89
   R. 'Europeana' 27, 28
   R. filipes 'Kiftsgate' 14
   Floribunda r. 87, 88, 120
   R. foetida 88
   R. 'Fragrant Cloud' 29
   R. 'Fru Dagmar Hastrup' 46
   R. gallica 46
   Gallica r. 46, 89
   R. gallica 'Versicolor' 46
   R. glauca (syn. *R. rubrifolia*) 89
   R. 'Geranium' 103
   R. 'Guinée' 55
   R. 'Hippolyte' 89

Hybrid Tea r. 87, 88
   R. 'Ispahan' 89
   R. 'Max Graf' 118
   R. 'Mermaid' 172
   R. 'Mme Hardy' 90
   R. moyesii 101, 103, 188
   R. 'Nozomi' 118
   R. × odorata 'Mutabilis' 89
   R. 'Paulii' 90
   R. 'Paul's Himalayan Musk' 29
   R. 'Peace' 27, 28
   R. 'Prima Ballerina' 27, 29
   R. 'Red Wonder' 27, 28
   R. 'Roseraie de l'Haÿ' 46
   Rugosa r. 46, 90
   R. rugosa 'Alba' 46
   r. rust disease 54
   r. replant disease 55
   R. 'Scabrosa' 46
   R. serafinoi 185
   shrub r. 85, 88
   R. 'Souvenir d'Alphonse Lavallée' 89
   R. 'The Queen Elizabeth' 46
   R. 'Tuscany' 46
   R. 'White Wings' 90
   R. wichurana 118
   R. 'Zéphirine Drouhin' 174
Rosaceae (rose family) 45–6
Rosmarinus (rosemary) 20, 24, 37, 117
   R. officinalis var. *angustissimus*
      'Benenden Blue' 117
Rudbeckia 112, 155, 157
   R. fulgida var. *sullivantii*
      'Goldsturm' 60, 179
rue SEE *Ruta*
rust disease 54
   rose r. 54
Ruta 22, 157

Sackville-West, Victoria (Vita) 81
Salix 41, 85, 187, 212
   S. alba 188

S.a. var. *sericea* 85, 188
    S. alba var. *vitellina* 'Britzensis'
      (syn. *S.a.* var. *v.* 'Chermesina') 41
    S. daphnoides 41
    S. exigua 85
    S. helvetica 85
    S. lanata 85, 187
    S. udensis 'Sekka' 41
Salvia 146
    S.interrupta 85
    S. lanata 101
    S. microphylla var. *microphylla*
      (syns *S. grahamii*, *S. microphylla*
      var. *neurepia*) 180
    S. patens 155
    scarlet s. 120, 156, 158
    S. × superba 143
Sambucus 40
    S. ebulus 94
    S. nigra subsp. *Canadensis*
      'Maxima' 85
    S. racemosa 'Plumosa Aurea'
      24, 40, 85
Santolina 57, 88, 121, 157, 188
    S. chamaecyparissus
      (syn. *S. incana*) 188
    S. pinnata subsp. *neapolitana*
      185, 188
Saxifraga stolonifera 111
    S.s. 'Cuscutiformis' 111
Sarcococca 110
    S. hookeriana var. *digyna* 110
Savill Gardens, Windsor Great Park,
    Berks. 122, 123, 165
Scabiosa caucasica 178
    S. graminifolia 178
Scarborough lily SEE *Cyrtanthus elatus*
Scrophularia aquatica 'Variegata' 26
Scutellaria canescens 89
sea holly SEE *Eryngium maritimum*
Sedum 'Herbstfreude' 86
    S. roseum SEE *Rhodiola rosea*

Senecio cineraria 22, 26, 57, 185,
    189, 190
Senecio greyi misapplied
    SEE *Brachyglottis* 'Sunshine'
S. laxifolius misapplied
    SEE *Brachyglottis* 'Sunshine'
Senecio reinholdii
    SEE *Brachyglottis rotundifolia*
S. viravira (syn. *S. leucostachys*)
    25, 133, 185, 189
Shasta daisy
    SEE *Leucanthemum × superbum*
Sheffield Park, E. Sussex 165, 197,
    198, 200
shrubs in mixed borders 82–7
    evergreen s. 146
Silene acaulis 201
silver leaf fungus
    SEE *Chondrostereum purpureum*
simazine herbicide 105
Sissinghurst Castle, Kent
    81, 90, 181, 185
Sisyrinchium angustifolium 153
Skimmia 21, 22
    S. japonica 43
Smith, James, of Tansley, Matlock,
    Derbys. (nursery) 64
Smyrnium perfoliatum 123
snapdragon SEE *Antirrhinum majus*
snowdrop 41, 99, *100*, 113, 114,
    204, 205
snowflake, summer
    SEE *Leucojum aestivum*
snowflake, autumn
    SEE *Acis autumnalis*
Solanaceae 181
Solanum crispum 25
    S. laxum (syn. *S. jasminoides*) 25
    S.l. 'Album' 180
sooty mould 64
Sophora japonica
    SEE *Styphnolobium japonicum*

sow thistle, creeping 50
Sowerby, John Edward,
    British Flora 199
Spartium junceum 100, *100,* 102
spear grass SEE *Aciphylla*
spearwort, lesser
    SEE *Ranunculus flammula*
speedwell SEE *Veronica*
    ivy-leaved s. SEE *Veronica hederifolia*
Spiraea 18, 30, 40
    S. japonica 'Anthony Waterer' *100*
    S. prunifolia 100
spleenwort, maidenhair
    SEE *Asplenium trichomanes*
spotted laurel SEE *Aucuba japonica*
spurge SEE *Euphorbia*
Stachys 157
    S. byzantina (syn. *S. lanata*) 154
Sternbergia lutea 143, 208
strawberry tree SEE *Arbutus unedo*
Street, John 33
Styphnolobium japonicum 53
sumach SEE *Rhus*
    Venetian s. SEE *Cotinus coggygria*
sunflower 157
    perennial s. 156, 157
sweet William 131, 132, 137, 138
    s.W. 'Morello' 137, 138
    s.W. 'Nigrescens' 138
sycamore 66, 153
Sycopsis 37
Syringa 38, 45, 57
    S. × *josiflexa* 'Bellicent' 39
    S. × *persica* 'Alba' 116
    S. × *swegiflexa* 'Fountain' 39
    S. vulgaris 24, 39
        S.v. 'Mme Lemoine' 39

Tagetes 'Cinnabar' 135
Tamarix (tamarisk) 97
    T. ramosissima (syn. *T. pentandra*) 97
        T.r. 'Pink Cascade' 97

Tamus communis 50
Tanacetum ptarmiciflorum 25
Taxus 20, 31, 32, 49, 55, 164
teazle 91
Thelypteris phegopteris
    SEE *Phegopteris connectilis*
thistle, creeping field 50
Thomas, Graham Stuart 204
Thuja 21
    T. plicata 33
thunder lily
    SEE *Zephyranthes grandiflora*
Thymus (thyme) 26
Tilia 68
    T. cordata 69
    T. × *europaea* 68
    T. 'Petiolaris' 69, 184
    T. tomentosa 69, 184
toadflax, ivy-leaved
    SEE *Cymbalaria muralis*
toetoe SEE *Cortaderia richardii*
Tompsett, Benjamin Philip 80
toothwort SEE *Lathraea clandestina*
top dressing 52
topiary 31
Tovey, Donald Francis 168
Treasure, Bernard John
    14, 88, 95, 170, 194
trees 14
tree mallow 59, 85
tree of heaven SEE *Ailanthus altissima*
Tritoma SEE *Kniphofia*
Tropaeolum speciosum 96
Tulipa (tulip) 58, 124, 158, 201, 205
    T. 'Dillenburg' 131, 140
    T. 'Dyanito' 121
    T. 'Golden Duchess' 133
    T. 'Niphetos' 127

Ulex europaeus 36
Ulmus 63, 64
    U. glabra 63

U. procera 63
Umbelliferae SEE *Apiaceae*

valerian, red SEE *Centranthus ruber*
Valley Gardens, Windsor Great Park,
    Berks. 65
vallota 52
    SEE ALSO *Cyrtanthus elatus*
van de Kaa, Romke 166, 167
Veratrum 57
Verbena 25, 156
    V. bonariensis 86, 91, 179
    V. rigida (syn. *V. venosa*) 179
Veronica hederifolia 58
veronica, shrubby SEE *Hebe*
Viburnum 24, 41
    V. × bodnantense 184
    V. farreri 184
    V. opulus 'Compactum' 41
        V.o. 'Roseum' (syn. *V.o.* 'Sterile')
            101, 103
    V. rhytidophyllum *101*, 103
Vinca 107
    V. difformis 98
Viola 127, 169
    V. biflora 204
    V. Clear Crystals Series 132
    V. 'Coeur d'Alsace' 203
    V. cornuta 140
        V.c. Alba Group 61, 177
    V. cucullata 114, 205
        V.c. 'Alba' 204
    V. 'Golden Dream' 136
    V. hederacea 205
    V. 'Lord Nelson' 127
    V. odorata 203
    V. pedata 169
    V. pubescens var. *eriocarpa*
        (syn. *V. pensylvanica*) 204
    V. riviniana 114, 204
        V.r. Purpurea Group 204

violet 91, 99, *100*, 114, 202–5
    common v. SEE *Viola riviniana*
    dog v. SEE *Viola riviniana*
    SEE ALSO *Viola*
virus disease 193

Wakehurst Place, Ardingly,
    West Sussex 193
wallflower SEE *Erysimum*
    Siberian w. 131
wand flower
    SEE *Dierama pulcherrimum*
weeds 58
Weigela 18, 30, 40, 57, 84
    W. florida 'Eva Rahke' 155
    W. florida 'Foliis Purpureis' 84
    W. 'Florida Variegata' 84
willow SEE *Salix*
    violet w. SEE *S. daphnoides*
    weeping w. 123
    woolly w. SEE *S. lanata*
willowherb SEE *Epilobium*
winter aconite SEE *Eranthis hiemalis*
wintersweet SEE *Chimonanthus praecox*
witch hazel SEE *Hamamelis*
    Chinese w. h. SEE *H. mollis*
Wolf, Hugo 200
Wright, Thomas 91

Xanthocyparis nootkatensis 34

yarrow SEE *Achillea*
yew SEE *Taxus*
York stone paving 151
Yucca 54, 88, 154, 166

Zantedeschia aethiopica 129, *129*
Zephyranthes 210
    Z. candida 142, 208, 210
    Z. grandiflora 210